S0-DTC-607

TEACHING RELATED SUBJECTS IN TRADE AND INDUSTRIAL AND TECHNICAL EDUCATION

TEACHING RELATED SUBJECTS IN TRADE AND INDUSTRIAL AND TECHNICAL EDUCATION

MILTON E. LARSON
Colorado State University

CHARLES E. MERRILL PUBLISHING CO.
A Bell & Howell Company
Columbus, Ohio

THE MERRILL SERIES IN CAREER PROGRAMS

Published by
Charles E. Merrill Publishing Co.
A Bell & Howell Company
Columbus, Ohio 43216

International Standard Book Number: 0–675–09073–3

Library of Congress Catalog Card Number: 72–80235

2 3 4 5 6 7 8–77 76 75
Printed in the United States of America

THE MERRILL SERIES IN CAREER PROGRAMS

In recent years our nation has literally rediscovered education. Concurrently, many nations are considering educational programs in revolutionary terms. They now realize that education is the responsible link between social needs and social improvement. While traditionally Americans have been committed to the ideal of the optimal development of each individual, there is increased public appreciation and support of the values and benefits of education in general, and vocational and technical education in particular. With occupational education's demonstrated capacity to contribute to economic growth and national well being, it is only natural that it has been given increased prominence and importance in this educational climate.

With the increased recognition that the true resources of a nation are its human resources, occupational education programs are considered a form of investment in human capital—an investment which provides comparatively high returns to both the individual and society.

The Merrill Series in Career Programs is designed to provide a broad range of educational materials to assist members of the profession in providing effective and efficient programs of occupational education which contribute to an individual's becoming both a contributing economic producer and a responsible member of society.

The series and its sub-series do not have a singular position or philosophy concerning the problems and alternatives in providing the broad range of offerings needed to prepare the nation's work force. Rather, authors are encouraged to develop and support independent positions and alternative strategies. A wide range of educational and occupational experiences and perspectives have been brought to bear through the Merrill Series in Career Programs National Editorial Board. These experiences, coupled with those of the authors, assure useful publications. I believe that this title, along with others in the series, will provide major assistance in further developing and extending viable educational programs to assist youth and adults in preparing for and furthering their careers.

Robert E. Taylor
Editorial Director
Series in Career Programs

FOREWORD

To teach the related subjects for an occupation is a challenge of major proportion. Either as a member of the occupational instructional team or as the only instructor in the program, the *must know* information of any occupation is essential to success.

A quotation, which is attributed to Benjamin Franklin, is "A trade without theory is like a tree without roots." The roots of any trade are mathematics, science, communications (graphic and verbal), and work adjustment knowledge. These must be mastered. It is to this goal that this book is directed.

The correlation of theory and skill does not come easy to the teaching-learning process. What is needed is a "team" effort if the relationship and proper student motivation are to be realized. How to correlate is the key to making the theory meaningful and to this objective the book will take the reader through the process of curriculum construction: preparing to teach, the teaching process, and evaluation.

The author, Milton E. Larson, is an outstanding teacher. He writes as he teaches: direct and to the point. As a leader in his field, his background includes a teacher at the post-secondary level; a director of trade and technical programs; and a teacher educator. His book is intended to "bridge the gap" for those who are sincere in producing keen minds and skilled hands to take their place in modern industry. To this end would be Dr. Larson's desire for all who read and study its content.

Carl J. Schaefer
Series Editor

PREFACE

The success of the student is closely related to the preparation of the teacher. Fortunately many teachers are endowed with great natural ability to teach. Most individuals can learn to be effective teachers and most have mastered the science and art of teaching through study of the teaching process and related activities.

Teaching related subjects in trade, industrial, and technical education is an important part of the preparation of students for jobs and positions in industry. The complexity of industry today with rapidly changing applications of technology demands much technical knowledge as well as technical skill of those individuals desiring entry-level jobs.

It is the hope that this book will help those who are teaching and those who are preparing to teach in vocational and technical education to be more effective. The book is designed to focus on major concerns of relevant teacher preparation and teacher performance in teaching those aspects of job information called related subjects. The essential information for the performance of the job or operation is often described as *must-know* information; while the information of less importance is classified as *nice-to-know* information.

This book is divided into five parts: building the performance base, construction of valid curriculum, preparation to teach, evaluation of performance, and the teaching process itself. Those who teach effectively in these five essential areas will provide most effective learning. It is with this purpose in mind that the author has identified the content and introduced the methodology contained in this volume. The concepts, illustrations, and supporting data are designed to help those who desire to teach to teach better. The author has tested many of the concepts in his own classes. Several of the exhibits were developed by students in the author's classes as part of the work of courses in analysis, curriculum development, and performance evaluation.

Milton E. Larson

CONTENTS

Part I BUILDING A PERFORMANCE BASE

Chapter 1 OCCUPATIONALLY ORIENTED STUDENTS AND LEARNING STYLES

The decade of the 1960s was the period of changing emphasis in education in the United States. Emphasis on "pay-check education" was reflected in federal legislation, expansion of state services, and unprecedented enrollment in local programs of vocational education. The voice of Congress was heard in every state and in most communities through the influence of the Vocational Education Act of 1963 and the Amendments to the Vocational Education Act, 1968. A new relationship of man, education, and work was recognized. More people realized that vocational education provided a bridge between man and his work, and enrollment data reflected a more realistic approach to preparation for life and earning a living. However, many still failed to realize that approximately eight out of every ten students entering grade school would not graduate from a higher education baccalaureate degree program.

The impact of unfilled jobs in technology and skilled occupations at the same time that high school graduates without salable skills were seeking jobs forced a "new look" at the total system of education in the United States. Riots in the streets added the voice of the disadvantaged and dissatisfied to those voices that were just beginning to be heard. Many asked the question, "Can our nation afford the luxury of a school system designed to meet the needs of a small percentage of the people while ignoring the needs of the majority?" Thinking citizens everywhere sought a new approach to the meaning of education. More and more the focus has been on vocational education as the solution to the problem.

Today's Students in Tomorrow's World

Today's students are the technicians and craftsmen of tomorrow. Industry's future depends on the availability of competent personnel. Automation and mechanization do not reduce manpower needs; they require instead that the manpower available possess a higher level of competency.

A gradual shift has occurred in the composition of our nation's work force. A nation of primarily blue-collar and agricultural workers has become predominantly composed of white-collar and service employees. This change has resulted from a heavy emphasis on technology. Personnel requirements have also gradually shifted to those occupations requiring longer training time and more advanced development of skills. The increasingly sophisticated nature of mechanized consumer goods has resulted in a rapid increase in the need for service personnel.

The *Manpower Report to the President* emphasized that:

> The pace of progress in developing this country's human resources will depend, in large measure, upon the educational system. Manpower development needs impose heavy new demands on educational institutions at every level. The experience of the past few years has demonstrated these institutions' ability, given needed resources, to take on new tasks and has shown the directions in which progress must continue.
>
> The schools' responsibilities for manpower development are threefold. One aspect is essentially remedial—to provide education and training for people who lack marketable skills or are employed below their capabilities, while job vacancies remain unfilled for lack of qualified workers. This is, of course, the objective of the training project for unemployed and underemployed set up under the Manpower Development and Training Act and other recent legislation, including the provisions for remedial basic education as well as skill development.
>
> A second aspect is to give young people still in school the best preparation for work and life. It involves education at all levels, from preschool to postgraduate, and in general subjects as well as

> those with a specific occupational orientation. It may also involve work experience, preferably integrated with schooling.
>
> A third major aspect is to provide for continuing education and updating of skills throughout working life. In part, the present need for training of the unemployed and underemployed reflects obsolescence of skills. But the problem is much broader than this. With the continued rapid pace of technological change and the mounting accumulation of knowledge, work preparation becomes a lifetime process, which educational institutions must aim to facilitate. (4, p. 75)

Three significant factors must be considered relative to the manpower problem and the education of youth. First, the total labor force; second, the change in distribution of employment by major occupations; and, third, youths' changing concept of their occupational objectives.

According to the *Manpower Report to the President,* by 1975 the country will have a working population of 154 million. (See Table 1–1.)

The nature of occupations and the distribution of employment by major occupational groups is changing. This change results in an increase of white-collar workers and service workers and a decrease of farm workers between 1947 and 1966, as shown in Figure 1–1.

In a study entitled *Vocational Education Interests of Colorado High School Students* in 1967, 75.39 percent indicated interest in vocational education. (See Figure 1–2.) This reflected the responses of 27,672 of the 36,705 students participating from 164 Colorado schools. The investigator found that this percentage compared favorably with the findings of a similar vocational survey conducted in Ohio which indicated that 72.4 percent were interested in vocational education. (10) A study made in selected schools in Utah in 1966 revealed that 83 percent of the students were interested in vocational education. While 75 percent of the high school students in Colorado desired the opportunity to take vocational education in high school, less than 10 percent of these students had the opportunity to enroll in programs that would prepare them for entry-level jobs. (See Figure 1–3.)

Students in the kinds of schools presently available will face a serious shock when they seek employment in tomorrow's world. This social problem was vividly portrayed by James A. Rhodes, former governor of Ohio, in his book, *Alternative to a Decadent Society,* when he said:

> Unemployment, welfare, and lack of skills are the major elements of a creeping paralysis that threatens our existence. Slowly but surely, as these elements multiply, they push society into a death grip from which there is no escape. Unemployment, welfare, and lack of skills are danger signals which tell us where society hurts. These signals must be answered by action. (6, p. 13)

TABLE 1–1

TOTAL LABOR FORCE, BY AGE, SEX, AND COLOR, 1955 AND 1965 AND PROJECTED 1975

(Numbers in thousands)

Age, sex, and color	Actual		Pro-jected 1975	Number change		Percent change	
	1955	1965		1955–65	1965–75	1955–65	1965–75
ALL CLASSES							
Both sexes							
16 years and over	67,988	77,178	92,182	9,190	15,004	13.5	19.4
16 to 24 years	11,668	15,653	21,061	3,985	5,408	34.2	34.5
16 to 19 years	4,637	6,353	7,865	1,716	1,512	37.0	23.8
20 to 24 years	7,031	9,300	13,196	2,269	3,896	32.3	41.9
25 years and over	56,320	61,525	71,121	5,205	9,596	9.2	15.6
Men							
16 years and over	47,405	50,946	59,355	3,541	8,409	7.5	16.5
16 to 24 years	7,483	9,758	12,995	2,275	3,237	30.4	33.2
16 to 19 years	2,908	3,833	4,664	925	831	31.8	21.7
20 to 24 years	4,575	5,925	8,331	1,350	2,406	29.5	40.6
25 years and over	39,922	41,188	46,360	1,266	5,172	3.2	12.6
Women							
16 years and over	20,583	26,232	32,827	5,649	6,595	27.4	25.1
16 to 24 years	4,185	5,895	8,066	1,710	2,171	40.9	36.8

16 to 19 years	1,729	2,520	3,201	791	681	45.7	27.0
20 to 24 years	2,456	3,375	4,865	919	1,490	37.4	44.1
25 years and over	16,398	20,336	24,761	3,938	4,425	24.0	21.8
NONWHITE							
Both sexes							
16 years and over	7,167	8,551	10,746	1,384	2,195	19.3	25.7
16 to 24 years	1,374	1,839	2,809	465	970	33.8	52.7
16 to 19 years	540	682	1,065	142	383	26.3	56.2
20 to 24 years	834	1,157	1,744	323	587	38.7	50.7
25 years and over	5,793	6,712	7,937	919	1,225	15.9	18.3
Men							
16 years and over	4,503	5,084	6,409	581	1,325	12.9	26.1
16 to 24 years	884	1,137	1,684	253	547	28.6	48.1
16 to 19 years	358	435	631	77	196	21.5	45.1
20 to 24 years	526	702	1,053	176	351	33.5	50.0
25 years and over	3,619	3,947	4,725	328	778	9.1	19.7
Women							
16 years and over	2,664	3,467	4,337	803	870	30.1	25.1
16 to 24 years	490	702	1,125	212	423	43.3	60.3
16 to 19 years	182	247	434	65	187	35.7	75.7
20 to 24 years	308	455	691	147	236	47.7	51.9
25 years and over	2,174	2,765	3,212	591	447	27.2	16.2

Note: Detail may not add to totals due to rounding.

Source: Manpower Report to the President. Washington, D.C.: Government Printing Office, 1969, p. 63.

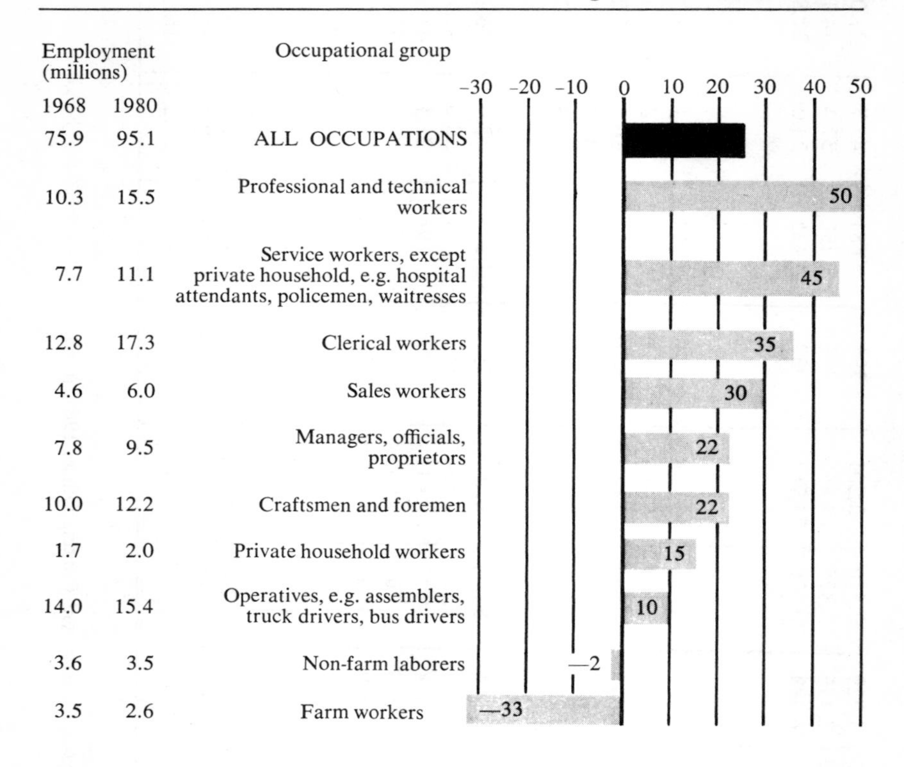

Source: U.S. Manpower in the 1970s—Opportunity and Challenge. Washington, D.C.: U.S. Department of Labor, 1970, p. 13.

FIGURE 1–1

PROJECTED GROWTH OF PROFESSIONAL, TECHNICAL, AND SERVICE OCCUPATIONS

Motivational Factors

Occupational motivation of youth as well as adults is dependent upon many factors and conditions. Why do youth make the occupational choices that they do? What influences their choices? How do their experiences aid the process of decision making? These and many other similar questions are basic concerns of teacher education.

Influence of the peer group, the family, and the experiences of youth contribute greatly to the individual's final occupational decision. Exposure at an early age to a large variety of occupations, both through

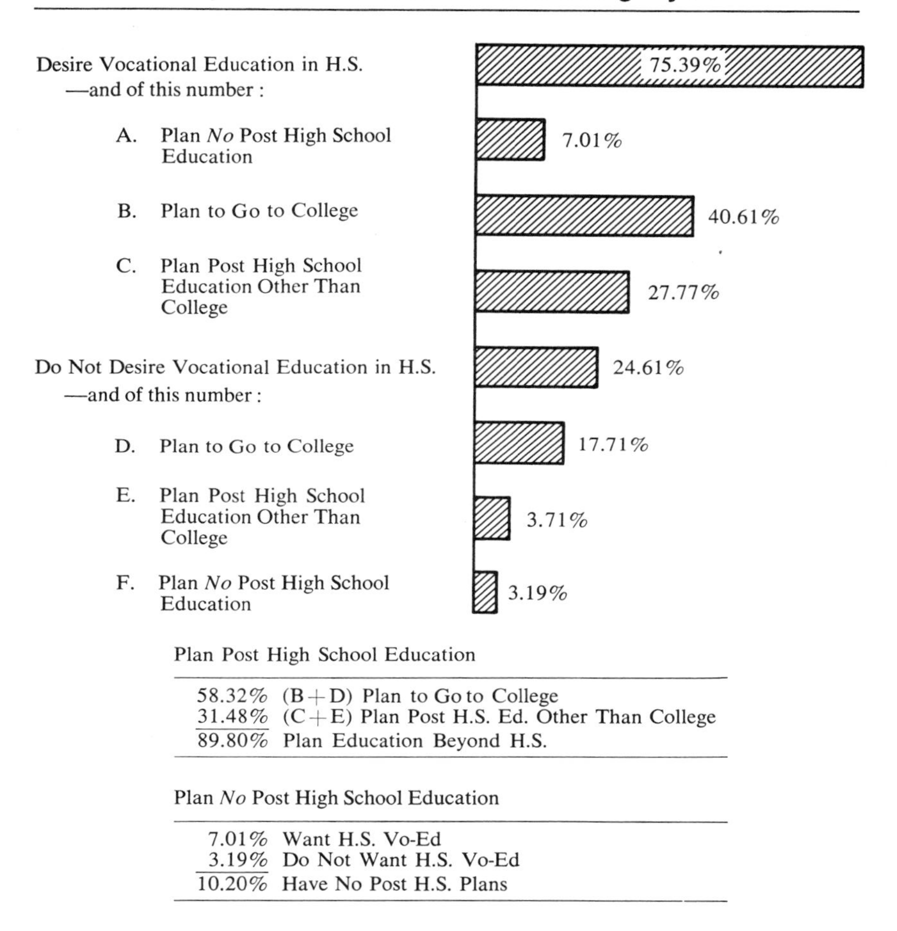

Source: Vocational Education Interests of Colorado High School Students, 1966–1967. Denver: State Board for Vocational Education, January 31, 1967, Appendix D.

FIGURE 1–2

COLORADO HIGH SCHOOL STUDENTS
DESIRING VOCATIONAL EDUCATION

vicarious experiences and through written and oral communications as well as media, aids the process of decision making.

Significant, too, is the orientation of the individual. Orientation is used to include the individual's desires, aptitudes, and abilities. In fact, the word is used in this book to include the sum total of all of the internal factors which affect the individual in his decision.

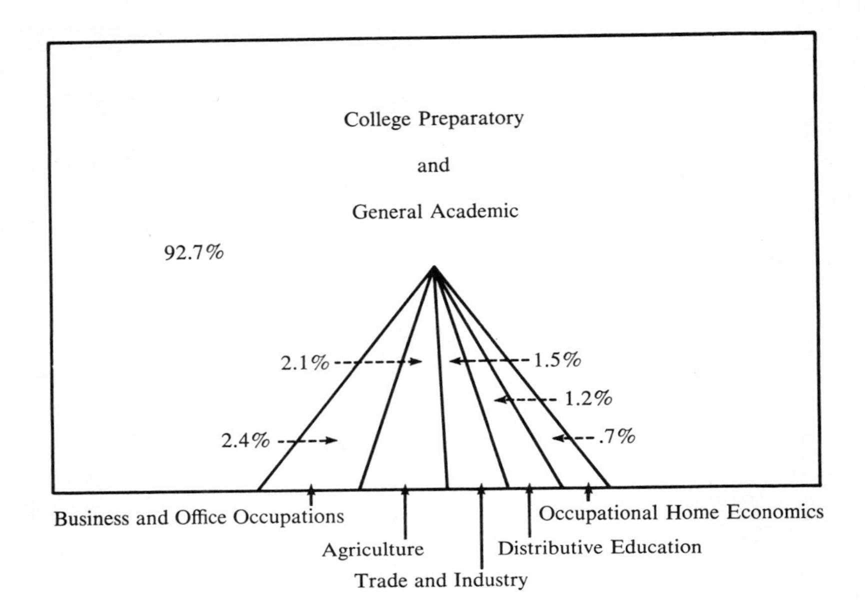

Source: Vocational Education Interests of Colorado High School Students, 1966–1967. Denver: State Board for Vocational Education, January 31, 1967, Appendix F.

FIGURE 1–3

HIGH SCHOOL ENROLLMENT
BY OCCUPATIONAL PROGRAMS

The motivation of each person is affected also by external factors such as:

1. Status of the occupation
2. Earning power of the vocation
3. Opportunities for promotion

Interest or motivation for many is more significant than intellectual superiority. Individuals with strong desires often excel in spite of limited intellectual endowment.

How does the teacher stimulate desire and increase motivation? Certainly, knowledge of the orientation of each student is a factor. Likewise, recognition of the opportunities for employment is equally important. Firsthand information of local placement opportunities and awareness of manpower requirements unmet in other parts of the nation are essential. Wage levels, fringe benefits, advancement opportunities, and concomitant job satisfactions are worthy factors for the student and his parents to consider in making decisions relative to job opportunities.

Emphasis on Real Situations

Three methods of learning are fundamental to teaching in vocational education. Many psychologists, including Skinner, have written about these methods. (8, pp. 7–8) Vocational educators usually relate these methods to learning through:

1. Experiences (informal)
2. Doing (planned job activities)
3. Trial and error (experimentation and research)

Vocational and technical education provide employmentlike situations in which students prepare for entry-level jobs. Much emphasis is placed on learning by doing both in shops and laboratories and in related classes. The conditions are similar to those found in industry, the problems are similar, and even the motivation for successful completion of the task or job is similar. While demonstrations and explanations precede shop and laboratory activities, some learning does take place by trial and error.

Experimental work involves an element of trial-and-error learning. The experiences of the learner and of the teacher are obvious as the student progresses from simple operations or tasks to more complex ones. The experiences, however, must be directed through a carefully designed instructional program to the achievement of the terminal behaviors identified at the time the goals were established.

Vocational and technical education have the advantage of occurring in a live laboratory where actual joblike experiences enhance the process of realistic everyday decision making as faced by individuals in the occupation. In such a situation the student has the best possible reason for learning—he is going to apply his knowledge and skills in performance which will result in some form of productivity. Often, students whose orientation fails to appreciate or recognize the advantages of academic educational activities alone readily recognize the possibilities of opportunities for satisfying careers in the many and varied fields of trade and industrial and technical education. Teachers have at their disposal a lifelike, joblike learning laboratory which gives meaning to abstract ideas.

Systems Designed for Individuals

Today, individually paced learning is to many people a new concept. To trade and industrial and technical educators this is not new. While the concept is not new, new "technologies of education" are being used with the old principle; therefore, many people see this as a new approach to learning and teaching.

Even though individuals in a class of students in trade and industrial and/or technical education are started together at the same level, in a very short time the more advanced students will have progressed beyond the average, while the slow students are behind the average. To effectively solve this situation vocational and technical teachers have used operation sheets, job sheets, information sheets, blueprints, and assignment sheets. These information and work sheets together with manuals, handbooks, and textbooks have for many years provided an effective vehicle for individualized or small group instruction. The advent of the workbook was a step forward in the instructional process. This process has now been advanced further with the arrival of programmed instruction, teaching machines, closed-loop film, and computerized instruction.

A systems approach is recognized as highly desirable. Such a system must include appropriate "software" as well as "hardware." Training programs in the armed forces have more and more developed a systems approach to the instructional program. Such an approach has been pioneered in vocational and technical education in the state of Washington.

Job-Related Concomitant Learnings

Sometimes academic teachers have been pleasantly surprised to find students who previously failed to master abstract concepts of English, mathematics, or other related subjects, succeeding in these areas after entering vocational and technical curriculums. Relevance is the "key" to educational achievement. When students realize the importance of reading or mathematics for electronics or auto mechanics or some other vocational field, then they learn the essentials of these subjects also. Experiences with the disadvantaged and students in the Job Corps have repeatedly affirmed this statement. In the words of Ornstein and Vairo:

> Disadvantaged children tend to like and do well in physical education, industrial arts, art and music; and teachers of these subjects have less difficulty with their pedagogy than others and find their efforts frequently rewarded. This is because the learning styles of slum children are best utilized in these areas: teaching is visual, concrete, and practical; it is physically oriented; it involves movement, excitement, and freedom of expression. (5, pp. 158–59)

Success-Designed Approaches to Learning

Success is a powerful motivating force. It is human nature to move away from experiences of failure and toward experiences of success.

Many opportunities exist in vocational and technical education for successful experiences. Vocational teachers who understand students help to provide experiences that are satisfying while at the same time providing the essential background for employment.

Students with beaming faces who have achieved success by meeting acceptable standards of quality and quantity of production are often heard to say, "I can do it!" For some students who have achieved meager success in academic classes this is a signal for celebration.

Effective demonstrations combined with meaningful explanations followed by individual instruction and increasingly challenging opportunities for mastery of problems generate confidence that leads to success. Success with humility combined with realistic goals keyed to performance on a payroll job are the result of effective vocational and technical education.

Career-Oriented Students' Goals

To comply with the provisions of existing federal acts providing funds for vocational education each vocational and technical student must have a vocational objective. Vocational education is not an exploratory or a recreational activity—it is strictly "geared" to preparation of students for jobs. This objective can only be achieved if an effective program of instruction is developed on the basis of careful analysis of the vocation or occupation for which the student is being prepared. Not only must the instruction be an outgrowth of the occupational analysis, it must also recognize the various learning styles of individual students.

Classes of Learning Styles

Torrance and White discussed the personal ways in which individuals process new information by identifying two cognitive learning styles. (9, pp. 239–40) One of these was concerned with conceptual tempos based on research conducted by Kagan while the second focused on selection strategies and involved the findings of Bruner, Goodnow, and Austin. The selection strategies as applied to learning concepts were described by Bruner and his associates as including one of the following: conservative focusing, focus gambling, simultaneous scanning, or successive scanning. Bruner pointed out that the individual may vary his stratagem with each new concept. Jerome Kagan related the learning style he identified to the socialization process in which he described some students as the impulsive type while others were considered members of the reflective group. He implied that the distinction was identified into the two categories as observed in relation to young children in problem-solving situations.

Marshall Rosenberg provides considerable insight into the learning styles of children. (7, pp. 18–30) He discusses specific learning skills as the aptitudes basic to learning in a classroom situation. Learning styles, on the other hand, are delineated by Rosenberg to refer to how these basic aptitudes are utilized in a classroom situation. The learning behavior of the student results from his learning skills plus his applied learning styles. Jensen related learning styles to individual differences. (2, p. 41) He considered learning styles to include those characteristics of the learner which have influenced the learner's performance.

Rosenberg discussed four learning styles:

1. Rigid-inhibited style
2. Undisciplined style
3. Acceptance-anxious style
4. Creative style

He advanced the theory that learning styles of individuals depend upon two dimensions of information-processing ability:

1. Locus of information
2. Level of symbolization

Locus of information was described as consisting of the degree to which the individual was open to receiving information from within himself and from sources outside himself. Rosenberg employed level of symbolization to describe the "level of abstraction with which the learner was able to symbolically manage information in a problem-solving situation." He recognized that considerable variation due to individual differences existed within each classification of learning style.

The characterization of these four learning styles as described by Rosenberg in his book, *Diagnostic Teaching,* follows:

> *Rigid-inhibited style.* When absolutistic principles are not available and the person is confronted with ambiguous or complex

problem-solving situations, his behavior is likely to become confused, disoriented, or withdrawn.

A learner of this type has trouble getting started unless directions are made simple and in concrete language.

Undisciplined style. Persons utilizing this style exert most of their energies in seeking immediate gratification of their needs and avoid behaviors which make them responsible to others.

As a rule, the undisciplined learner performs better on tasks which allow him to perform autonomously than he does on tasks which require his following the directions of others.

Acceptance-anxious style. The acceptance-anxious learner tends to react catastrophically to errors and criticism. Rather than seeing them as a way to improve upon his learning, he sees them as a rejection of his worth as an individual.

The acceptance-anxious learner is more comfortable in a passive learning situation wherein the teacher presents the "right" answer.

Creative style. The creative learner can be primarily differentiated from learners using other styles by his flexibility and openness. He perceives errors as opportunities to learn how to improve upon his performance. He thus is not frightened, guilty, or defensive when errors are made. He persists on tasks, not out of fear, but out of intrinsic interest in gaining a sense of mastery over the problem.

The psychologists mentioned did not identify specific learning styles with any occupational group. Without doubt all of the learning styles are represented among the students entering programs of vocational and technical education. Teachers and other educators in this field need to be strongly aware of the differences between learning skills and the styles of learning employed by boys and girls.

After further research the learning styles of students who select vocational and technical education may be delineated. The author believes that these students possess a strong orientation which has been a major factor in the decision to enter vocational and technical education. It may well be that in a large number of individuals a specific learning style can be identified. If this hypothesis should ever be positively affirmed, a valuable tool would be available to use together with knowledge of learning skills in better identifying, counseling, and educating these individuals in trade and industrial and technical occupations.

Dynamic Teaching

In the words of Charles Jones, "The purpose of education is to change behavior." (3, p. 1) This is the purpose of all vocational and technical education. The changes of behavior must be measured in terms of preparation for employment in the field studied.

Dynamic teaching grows out of a strong dedication to the field plus a positive conviction that others can acquire fundamental competencies essential to success.

The worth of the task has long since been accepted as essential to the success of individuals, the growth of industry, the progress of society, and the survival of the nation. The validity of this concept has been tested in time of peace as well as in time of war.

Refinement of a systems approach to vocational and technical education is now essential if the 80 percent who do not earn four-year college degrees are to be prepared for entry-level jobs in a rapidly changing technological society. The systems approach must contain a major emphasis on a validated curriculum based on analytical study of occupational activities. Such occupational analysis can provide a solid foundation for development of performance goals, instructional content and systems, and meaningful evaluation.

Integrated into the systems approach must come a new orientation toward people. Bugelski identified this as human engineering. (1, p. 9) He related this to the industrial engineering approach as used by Franklin W. Taylor. Human engineering concentrates on an individual's goals and examines these in relation to the work environment.

The methodology of human engineering may lead to the solution of many problems of deep concern to vocational and technical education.

Summary

Vocational and technical education is being regarded by more and more people as essential to prepare the majority of young people for productive and creative service.

Large percentages of students want opportunities to acquire vocational and technical competencies. These competencies are acquired in many different places in many different ways.

What people learn is constant, but how people master the knowledge and skills varies. Different people have different learning styles. The orientation of individuals is expressed through the individual's capacities, interests, abilities, desires, aptitudes, and previous experiences.

The orientation of an individual influences his learning style.

QUESTIONS AND ACTIVITIES

1. Discuss the relationship of man, education, and work.
2. How has recent federal legislation modified the traditional concepts of vocational education?
3. Who is responsible for matching people with jobs? How?
4. What percentage of students in high school should be in vocational and technical education? Of students in post-high school institutions? Why?
5. Is "paycheck education" a good term to use? Why?
6. What provides motivation for the occupational choices of youth?
7. Discuss the status of occupations as conceived by children, youths, and adults.
8. What is meant by learning styles?
9. What is the effect of success upon motivation of the individual?
10. Discuss the decision-making process as related to choice of careers.
11. Compare the learning styles of students you have known.
12. Are certain learning styles characteristic of students who seek vocational and technical education? Explain.
13. How can teachers profit from knowledge of learning styles?
14. What are the implications of learning styles for:
    a. Program planning?
    b. "Technologies" of education?
    c. Guidance and counseling?
    d. Administration and supervision?
    e. Facilities planning?
15. Explain what is meant by a systems approach to vocational and technical education.
16. Delineate the role of the teacher in the modern process of vocational and technical education.

REFERENCES

1. Bugelski, B. R., *The Psychology of Learning Applied to Teaching.* Indianapolis: The Bobbs-Merrill Co., Inc., 1964.
2. Jensen, A. R., "Varieties of Individual Differences in Learning," in *Learning and Individual Differences,* edited by R. M. Gagné. Columbus, Ohio: Charles E. Merrill Publishing Company, 1967.

3. Jones, J. Charles, *Learning*. New York. Harcourt, Brace & World, Inc., 1967.

4. *Manpower Report to the President*. Washington, D.C.: U.S. Department of Labor, 1969.

5. Ornstein, Allan C., and Philip D. Vairo, *How to Teach Disadvantaged Youth*. New York: David McKay Co., Inc., 1969.

6. Rhodes, James A., *Alternative to a Decadent Society*. Indianapolis: Howard W. Sams & Co., 1969.

7. Rosenberg, Marshall B., *Diagnostic Teaching*. Seattle: Special Child Publications, 1968.

8. Skinner, B. F., *The Technology of Teaching*. New York: Appleton-Century-Crofts, 1968.

9. Torrance, E. Paul, and William F. White, eds., *Issues and Advances in Educational Psychology: A Book of Readings*. Itasca, Ill.: F. E. Peacock Publishers, Inc., 1969.

10. *Vocational Education Interests of Colorado High School Students, 1966–1967*. Denver: State Board of Vocational Education, January 31, 1967.

Chapter 2 STRUCTURING LEARNING TO RESULT IN PAYCHECK JOBS

"Education with a mission" should be the goal of all school experiences for all individuals. One of the most important missions is to provide that kind of education which results in productive, useful, and responsible citizens. Education must be realistic from the point of view of the learner as well as from the point of view of the teacher. During 1969, Vice President Spiro Agnew emphasized repeatedly that vocational and technical training is too often being neglected in favor of "the elegant ornament of the liberal arts."

In the first annual report of the National Advisory Council on Vocational Education (July 15, 1969) it was stated that not only unemployment, but also much of the unrest and violence in our cities, was the result of the failure of our schools to produce individuals prepared for employment. The report further stated that the education of nearly twenty-five percent of the young men and women who turn eighteen each year is a waste of money as well as human resources. The report continued with a statement that "schools can prepare young people to realize their potential." The reasons for the high failure rate were identified as follows:

1. the attitude that vocational education is for someone else or somebody else's children

2. the lack of adequate programs
3. insufficient money for vocational and technical education (1)

Attitude, program, and money are the three vital elements for an effective vocational and technical education for both youths and adults. The concern in this chapter is the lack of adequate programs.

Subjects Significant for Vocational and Technical Education

Vocational and technical education cannot result in paycheck jobs unless proper balance is maintained among general education, education in the specialty subjects, and education in the related subjects.

Occupational choice and the decision-making process are two subjects of much concern to those seeking to improve paycheck education. Many psychologists have agreed that the process of decision making can be expedited by providing students with experiences which supply adequate resources for making occupational decisions. Visiting places of employment, viewing films and other audiovisual aids, and reading books and articles—all of these expose the students to a broad variety of work experiences and hasten the development of their ability to make realistic decisions. Identification of the role of work and respect for work itself in our form of society adds to the motivational forces which lead to realistic decision making.

During the early grades the emphasis should be on broad families of occupations. As the students progress the focus is sharpened. It is not to be expected that initial decisions will not be changed or modified. This usually takes place as students mature and fortify the body of their own knowledge and relate to the desires, abilities, and interests in the world of work. The process of decision making relative to occupational choice is for many a very gradual, evolving type of experience. Tryout experiences, in varying degrees depending upon opportunities, strengthen or nullify the early concepts of children and youth. As youth progress into senior high school, cooperative employment, part-time work experience, and summer employment are helpful in tempering previously conceived ideas of likes and dislikes, abilities and handicaps, as well as dreams and ambitions.

The school can do much to assist students in the exploratory stages of growth and occupational development. These are responsibilities of general education and prevocational education. Attention to these opportunities for educational development needs to be centered during elementary and junior high school years. True, some "late bloomers" will not reach the stage of maturity and self-concept reality essential to making such decisions until senior high school or after completion of high school. The sooner identification can be made with the occupational world the more likely education will take on a meaningful experience for the student. Our nation can no longer afford the waste of meaningless education or education for goals that are unsought, incapable of achievement, and out of harmony with the goals of students, employers, and the nation.

The Specialty Subjects

The specialty subjects provide the concentration of subject matter knowledge and skill essential to perform as an employee on the payroll job. This is the heart of any vocational and technical training program. It is the focus of the educational goal. Without essential competencies acquired through the specialty subjects the student would have nothing to sell that most prospective employers desire in a new employee. Most employers expect all candidates for employment, except novices, to have enough ability to perform certain tasks, operations, and jobs assigned with minimum assistance and supervision. This is, in effect, what some employers are saying when they insist on experienced candidates for employment. They are saying that they desire individuals who have demonstrated the ability to perform on the paycheck job under similar conditions. Other employers are insisting on individuals who have completed programs of instruction in specific institutions as a substitute for experience. Again, these employers are saying that these individuals are likely to have acquired the competencies which they desire and therefore are likely to perform acceptably on the job.

The specialty subjects are usually concentrated in the shop and/or in the laboratory. If the program is machine shop, auto mechanics, welding, carpentry, electricity, etc., the specialty subjects are the shop courses. If the program is mechanical technology, electronics, electromechanical, instrumentation, etc., the specialty courses are those shop and laboratory courses that prepare the student to perform psychomotor plus cognitive functions on the payroll job.

The Related Subjects

To expedite the instructional programs the curriculum is subdivided into courses. Often the courses are broken down into units of instruction. Each unit in turn is divided into smaller parts sometimes referred to as lessons.

The courses of the specialty are often separated from related courses. This may result from the division of the instructional work load between or among teachers. To utilize the expertise of the shop or laboratory teacher to the maximum in his specialty necessitates other teachers' providing the related instruction.

Using a circle diagram of the total instructional programs (see Figure 2–1) places the specialty subjects at the nucleus of the circle or the focus

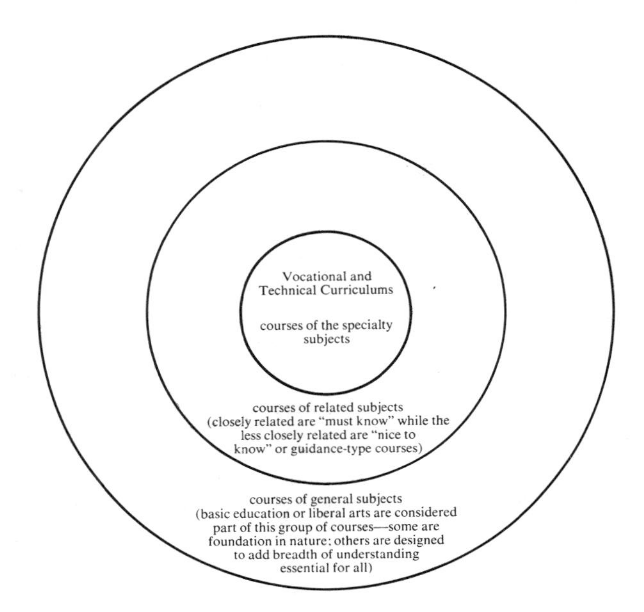

FIGURE 2–1

STRUCTURE OF SUBJECT AREAS IN VOCATIONAL AND TECHNICAL EDUCATION PROGRAMS

of the program. The second zone contains those courses which are related to the specialty. Those related courses which the student depends upon in order to perform the tasks of the specialty may be called "must know" related subjects. Other related subjects which are less essential to the performance of the tasks but are helpful or provide information of a guidance nature are referred to as "nice to know" subjects.

In some instructional programs all related information is grouped into a related class. Under this type of organization the content of the related course may be divided into "must know" and "nice to know" information.

The modular system of structuring instructional programs provides a plan of organization which lends itself to a division of responsibilities between or among instructors of a particular curriculum. Under the modular plan some classes, such as the shop or laboratory period, may consist of three modules, while the related class and the general education class each consists of one module. The size of the module may be fifteen minutes or twenty minutes instead of fifty or fifty-five minutes. Scheduling may be in terms of consistent numbers of modules for each class throughout the year or quarter. In some institutions the number of modules used for each subject varies from day to day in accordance with the decisions made by the instructors at the previous planning sessions. Adjustment of numbers of modules from day to day is usually accomplished by the use of a computer. Whether or not a sophisticated system of scheduling is employed, related subjects are often separated from the specialty subjects. However, in some schools the related information is provided by the same instructor during the same period as the specialty subject. This system is usually identified as "over-the-shoulder" teaching of related instruction. This practice is more common in secondary than in post-secondary schools. It is used more in trade and industrial curriculums than in technical programs.

Courses of related instruction should be closely correlated with the specialty subjects. Ideally, the concepts of information relative to the new topic to be studied in the shop or laboratory should be presented in the related class prior to the time it will be needed in the shop or laboratory. Principles of learning and teaching should be recognized as vital to the success of the related instruction. Some of the most significant principles are:

1. The learner views new opportunities and content through previous personal experiences and acquired knowledge.
2. Retention of skills and knowledges depends upon use, competency, and confidence, while efficiency depends upon frequency of application.
3. Learning is dependent upon the relevance of the learning experience.
4. Learning which brings pleasant results is most effective.
5. Reinforcement of learning is essential for most effective performance. Overlearning is a method of reinforcement.
6. Involvement in concrete experiences is the most effective way of learning. All learning must grow out of that which the learner already knows.

All related subjects may be grouped into two classifications: generally related and technically related.

GENERAL EDUCATION RELATED COURSES

For purposes of this book basic or fundamental education and liberal arts education are considered as general education. Many people view general education as a forerunner of specialty education. Certainly the ability to read and interpret the written word as well as the application of some mathematics is necessary for most, if not all, fields of vocational and technical education even at the most primitive stages and beginning levels as applied in the twentieth century.

Some students lack motivation for general education. One of the biggest challenges to education today is to make education relevant. This is especially true of general education and liberal arts since the objectives of the courses often are not in harmony with the interests of the learners. Experiences with disadvantaged students and students in Job Corps camps have testified to the significance of motivation and relevancy as related to student learning. When students realized that the ability to read and interpret a manufacturer's manual was vital to the repair of a mechanical unit, they gained strong motivation for learning to read. When the application of mathematics was identified as necessary to the solution of a problem in mechanics or design, the learner was successful in mastering concepts and processes which previously had been meaningless.

Should basic education precede or should it occur simultaneously with experiences in vocational and technical education? This may be a question to consider relative to the needs of those who have been exposed to several years of elementary and high school without really acquiring adequate competencies in general education to perform successfully in specialty subjects. Another factor to consider may be the introduction of broad elements of vocational and technical education (often referred to as career education) at an early age in order to give greater relevance to subjects in general education.

General education provides the tool subjects for expediting mastery of the specialty subjects. The more complete the set of tool subjects, the easier the mastery of the specialty subjects. In fact, from the point of view of vocational and technical education, the subjects of general education may be considered as providing a service function.

General education related courses include English composition, general mathematics, social science, and humanities. The title of courses in each of the above areas will vary with the level of the educational program. At the high school level much of the general education emphasis is on English, history, or other social science, and on such mathematics as algebra, trigonometry, and geometry. At the post-high school level combinations of subjects may be grouped more broadly.

General education subjects are usually organized into departments devoted totally to this field. Teachers specialize in a particular branch of this field such as English, mathematics, social science, etc. Most of these teachers have an academic background with little or no payroll experience outside of teaching. This often results in serious problems relative to making the courses of general education meaningful to students who have an orientation toward vocational and/or technical education. To teach students in vocational and technical education general subjects successfully, teachers must be able to translate fundamental principles and theories into meaningful applications and solutions to problems. To achieve this objective teachers must not only know the subject matter, but must also have some grasp of occupational fields. Illustrations, problems, and experiences drawn from the world of work provide meaning and motivation to students. A close working relationship between the teachers of general education subjects and the teachers of specialty subjects is essential if students are to utilize effectively the content of both series of courses. This problem has become intensified due to the splintering of subjects and the combination of students from a number of different vocational and technical fields into the same sections of the general education courses. The problem is further complicated by the inclusion of students other than vocational and technical students in the same classes.

Courses in general education are usually identified with such goals as:

1. Understanding of democracy
2. Appreciation of the heritage and culture of our society
3. Mastery of basic skills of communication
4. Ability to live and work with others
5. Ability to perform basic mathematical calculations
6. Development of worthy use of leisure time

These are all highly commendable goals which should be sought by each person and by each citizen of our nation. But today, goals of general education must be related to occupational goals whether these goals are achieved immediately upon graduation from high school or are of a post-secondary nature.

How should the content of courses of general education be derived? What content will ensure the achievement of identifiable goals? How can the goals be translated into behavior reflecting achievement? Are behavioral objectives determined for each of the courses of the general education sequence?

Vital teaching content results only from careful analysis of the subject and development of behavioral objectives. Subject matter content must be adequate to provide the learner with that knowledge requisite to the goals. Evaluation must be in terms of the established objectives. Most courses in subjects of general education would become more relevant if the program planning provided adequate emphasis on three basic steps:

1. Analysis of the subject
2. Development of behavioral objectives
3. Evaluation in terms of behavioral objectives

Program planning and curriculum building, while somewhat more elusive in general education courses than in technically related courses, are just as important. Dropouts and failures in high schools are clear indications of the challenges still to be met in this vital area. Integration of resources through curriculum centers is one way to move more rapidly in the direction of meaningful content for students in academic general education courses. Better teaching materials, reinforced by teachers more aware of the needs of today's students and having themselves closer contact with the world of work, are other goals to be sought to strengthen general education courses.

Methods of teaching and learning must strengthen curiosity, stimulate interest, develop motivation, and at the same time recognize individual differences. Flexible organizational structures are helpful in achieving a good learning and teaching situation. Individually paced learning, small group instruction, and large group instruction are different methods which should be selected and used when most appropriate to achieve the established goal. Class participation, problem solving, and simulated instruction are all teaching processes that when skillfully employed provide teaching for maximum meaningful learning.

Recognition of the orientation of vocational and technical students is necessary if teachers of academic general education courses are to build a bridge from the reality of the vocational world to the fantasy of many situations existing in general education classes.

TECHNICALLY RELATED COURSES

Technically related courses for the vocational or technical specialty are in this book used to include the courses sometimes referred to as applied subjects. The distinction between related and applied subjects is one of organization of classes more than of derivation of subject matter content.

Technically related courses include those courses that directly provide supporting knowledge or information essential to the performance of the specialty. These may be related science, related drawing, related mathematics, etc. The content of each course is delimited to that part of the subject which relates closely to the specialty. No attempt is made to treat the entire subject. Each course provides background information to strengthen the competencies of the learner.

Intensity of the objectives and available time are factors in the determination of the amount of "nice to know" content. First, the "must know" content must be thoroughly mastered by the student. Without mastery of this, performance of the job would be either impossible or ill-advised.

While in some courses of the specialty related instruction is limited to "over-the-shoulder" instruction, the common practice is to provide one or more classes of related instruction each day to accompany the shop or laboratory classes. These are frequently structured as a fifty or fifty-five minute period. Some of the related courses are frequently taught by the shop or laboratory teacher; this is rarely the case with applied subjects in community colleges.

The success of the related subjects courses is greatly affected by the teacher's ability to integrate the course content with the needs of the student in the specialty subject. Timing is important. The topic is most effectively taught just prior to the time that it will be used in the specialty subject. The advantage in having the shop or laboratory teacher teaching the related subjects is apparent in that he then can regulate the emphasis and timing to meet this need. In situations where other teachers have responsibility for the related subjects course, close correlation between the shop and laboratory instructors and the related subjects instructors is very important to the students' achievement in both series of courses.

The content of related courses should result from occupational analysis. Identification of content and priorities of content will be gained through such analysis. (See Chapter 4.) Once the content has been identified and the priorities established the integration into the instructional program involves understanding of the learning process.

The core-cluster approach is used where several vocational and technical curriculums need the same related courses and the order of the content can be adjusted to meet the needs of the students in their various specialities. For example, students in machine shop and in sheet metal may be grouped in a core-cluster course in blueprint reading and another in related science. Students in automotive body repair may be grouped with those in sheet metal in a related course in metallurgy. More and more schools are using the core-cluster approach. This is especially true where the enrollment is too small to support an adequate class in a specific related subject. Another advantage of the core-cluster approach lies in the better use of facilities and instructional materials. Likewise, specialization of instructors is a factor in the decision to follow the core-cluster approach.

The main disadvantage of such an arrangement lies in groupings that do not really lend themselves to clustering—this may actually interfere with learning instead of expediting it. Again, the instructor must have adequate background to be able to relate to the students' interests and needs in his field of specialty.

The rapid growth of industry and the new innovations of technology have resulted in a situation which demands of each tradesman and technician a more comprehensive and thorough understanding of the science and technology involved in the performance of the task. This is true of service functions as well as production and experimental functions. As a result, more emphasis and time must be devoted to related subjects. The more advanced the industrial application becomes, the more related information is required by the employee. This is indicated in Figure 2–2, which illustrates the amount of technical skill (knowledge) required by the skilled craftsman and the technician as compared with the semiskilled worker. Likewise, as the amount of technical knowledge required increases, the time devoted to manipulative skill decreases. This is also shown in Figure 2–3, which illustrates the types of technical occupations below the professional level. Figure 2–3 uses the word "level" to indicate the depth of knowledge and skill required and the word "scope" to convey the concept of the breadth of knowledge and skill necessary for the performance of the job classification indicated.

Maximum learning and opportunities for advancement are both greatly affected by the nature and content of the courses in the technically related subjects. The "how" is important and is part of this content, but

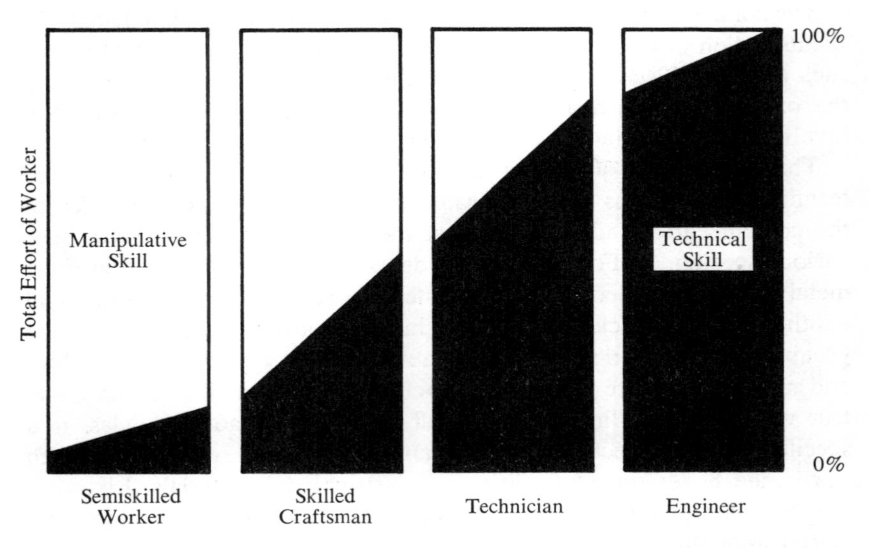

Source: Lynn A. Emerson, *Technician Training Beyond the High School.* Raleigh: Vocational Materials Laboratory, State Department of Public Instruction, June, 1962, p. 8.

FIGURE 2–2

TECHNICAL AND MANIPULATIVE SKILLS OF WORKERS IN INDUSTRY

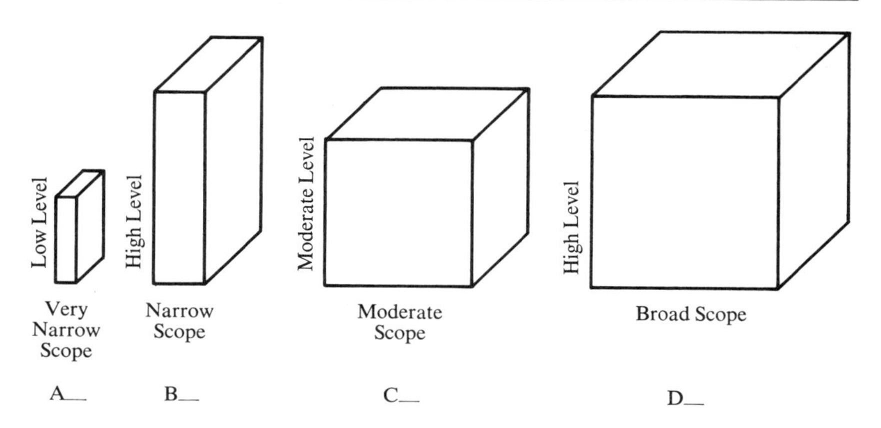

A. Low-level narrow-scope technical occupation, not classified in the technician group.
B. Technical Specialist)
C. Industrial Technician) May be classified as technician occupations.
D. Engineering Technician)

Source: Lynn A. Emerson, *Technician Training Beyond the High School.* Raleigh: Vocational Materials Laboratory, State Department of Public Instruction, June, 1962, p. 10.

FIGURE 2–3

SCOPE AND LEVEL OF WORKERS IN INDUSTRY

the "why" must not be ignored if the growth of the student in his specialty is to continue.

Summary

Education must be relevant if it is to be accepted by students in today's world. This is true of general education, related education, and the specialty subjects. Relevance motivates learning and expedites teaching. It results from building a bridge between the world of work and the world of school.

Analysis of the world of work provides a realistic foundation for all courses related to vocational and technical occupations—general, related, and specialty subjects.

Time dictates the need for priorities in the selection of content for each course. Most vital to success on most vocational and technical jobs in *decreasing* order are the:

1. Specialty subjects
2. Related subjects
3. General subjects

The related subjects should grow out of and support the specialty subjects. Application of the principles of learning enhances learning in related as well as in all classes.

Most effective learning results when general and related course content immediately precedes and is integrated with the topics of the specialty subjects being studied.

The content of related education should:

1. Result from occupational analysis.
2. Employ behavioral objectives.
3. Evaluate performance in terms of identified behavioral objectives.

A more intensive study of more complex related instructional content is essential for employees in today's highly technological society. The "why" is important together with the "how" if the employee is to advance in today's competitive world.

QUESTIONS AND ACTIVITIES

1. Discuss the relationship of school programs to social unrest.
2. Do you agree that "vocational education is for somebody else's children"?
3. What can vocational and technical educators do to improve the attitude toward vocational and technical education?
4. Discuss the implications of the criticisms given by the National Advisory Council on Vocational Education in its first report.
5. Should individuals be encouraged to explore potential occupations while still in junior high school? Discuss.
6. How may individuals be better prepared for making decisions relative to occupational choice?
7. What is the function in programs of vocational and technical education of subjects of general education, related education, specialty education?
8. What is the difference between "nice to know" and "must know" related educational contents?
9. What are the implications of modular scheduling for vocational and technical education?
10. How can related instruction be made most relevant for students?
11. Discuss the implications for related course teachers of the following:
    a. "All learning must grow out of that which the learner already knows."

b. "Retention depends upon use."
c. "Learning is dependent upon the relevance of the learning experience."

12. How do generally related courses differ from technically related courses?
13. How can the "focus" of objectives be sharpened for general education? For related education?
14. How do teachers of related courses benefit from occupational analysis?
15. Are behavioral objectives applicable to the related courses? Justify your position.
16. "Timing" of content of related courses is important. Why?
17. Is the base of technical education science or is it mathematics? Why?
18. How can general and related courses be made more meaningful to students?
19. What competencies should instructors of related courses possess?
20. Why are general education courses important for vocational and technical students? Related courses?

REFERENCES

1. "First Annual Report of the National Advisory Council on Vocational Education." Washington, D.C.: Office of Education, U.S. Department of Health, Education, and Welfare.

Chapter 3 INTEGRATING SUBSTANTIVE CONTENT INTO RELATED INSTRUCTIONAL MATERIALS

Teachers of related subjects have vital functions to perform in the plan of education of students in trade and industrial and technical education. Related subjects may be taught by the same teacher who is teaching the shop or laboratory subject, but frequently it is another teacher who has been given the assignment of teaching one or more courses of related mathematics, science, blueprint reading, drawing, etc.

The increasing complexity of technology places greater demands upon the related courses. These courses are now more essential than ever. It is more vital than ever that individuals teaching these courses present the content in a relevant and effective manner to enhance the climate for students' learning.

Frequently, students have difficulty performing effectively in the shop or laboratory because the content of the related courses is meaningless. Sometimes the content is selected with little attention to its relationship to the job or its value to the students. For this reason individuals selected to teach related subjects must relate to both the students and the shop or laboratory teachers.

Course content must be selected in accordance with priorities cooperatively determined by the shop or laboratory teachers. Useful in making such determinations is the advisory committee for the trade or technology. The occupational and task analysis provides positive indicators of the importance and the priorities of subject matter content for units or courses of related instruction.

Course Content for Related Instruction

The purpose of related instruction is to provide a more effective base of operations for the student in school and later on the job. Priorities must be placed on the content most essential for the students to know.

Three kinds of content are frequently grouped together as related instructional materials: *must-know, nice-to-know,* and *guidance materials.* As the name implies, the *must-know* related information is vital to the students' success in school and later on the job. The *nice-to-know* content adds some enrichment but is not absolutely essential to success either in school or on the job. *Guidance materials* aid the students in making decisions relative to the occupational choices available as to specialties, job opportunities, working conditions, opportunities for advancement, and so on.

The process of content selection for related instructional materials should be made in the same manner that the content is selected for the specialty of the shop or laboratory. More and more emphasis is placed today on the performance base. This means that the content should relate to the needs for knowledge identified as essential on the job. The best way to make such a determination is to study the work functions performed on the job. Having identified the "doing" elements, it is then necessary to determine the related "knowing" content. This is most effectively done by an occupational and job (task) analysis. If such an analysis has not been performed or is not contemplated, then heavy reliance on the opinions of the advisory committee or other consultants is a sound alternative. Failure to employ either of these frequently results in selection of related instructional materials lacking in relevance and unsuited to the purposes for which it is intended.

It is a faulty assumption to say, "Everyone needs to understand mathematics, science, English, etc." Such generalizations lead to the inclusion of much content which is irrelevant and unimportant for the student in vocational and technical education. Such an approach to determination of content increases the tasks of the shop or laboratory teacher and often drives students out of school. There is little holding power in meaningless content or content taught in such a way as not to relate to the goals of the students and his purposes in seeking an education.

Course content for related instruction should be determined by a cooperative effort of the shop and laboratory teachers with the related instructors utilizing a valid performance-based criterion.

Teachers of Related Instruction

Who should teach related courses? What qualifications should he possess? Is a generalist best for such an assignment? Should courses in related mathematics be taught by instructors from the mathematics department who teach the theory of mathematics? Should related instructors possess some occupational experience? These and many other questions of a similar nature occur to people faced with the problem of selecting teachers for related courses. In some cases, requirements of the state for credentialing purposes will automatically answer some of these questions, otherwise reimbursement for the vocational related teacher will not be forthcoming.

Some believe the shop or laboratory instructor is the best person to teach the related subjects as well as the specialty subject of the curriculum. Individuals who take this point of view often express the concept that this teacher "knows best" what the students need and the content can be taught at the time when it best fits into the program of instruction. Timing is important—the topic may be essential for tasks to be performed in the shop or laboratory. This is especially true of mathematics and science. Teaching how to perform a certain type of calculation just before the student needs to apply the principles to a problem which he must solve in connection with a shop or laboratory assignment is the most effective time from both the students' and the instructors' point of view. This also provides effective motivation and relates the work of the two classes. Such timing rarely occurs by accident. Careful planning and close cooperation when two or more instructors are involved is necessary to achieve a high degree of coordination. However, if one person is responsible for the content of both subject matter fields this sort of problem vanishes. In small vocational schools this philosophy quite frequently is mandated by having only one teacher. It is, however, much more common to have instructors other than the shop or lab teachers teach the related or applied courses.

If special instructors are assigned to teach the related courses, what qualifications should these instructors possess? Obviously, the following are important:

1. Competency in the subject matter.
2. The ability to teach the content.
3. Recognition of the applications of the content to the goals of the students studying the occupational program.
4. The ability to see their role in perspective to the role of the teachers of the occupational specialty.
5. Empathy for vocational and technical students.

Individuals who teach related subjects must be qualified in the subject matter. The theory is essential, but it must be related to the applications which students will need to make in the shop or laboratory and later in the world of work. Vocational students are not interested in learning theory for theory's sake. The theory must become fundamental principles that explain the "how" and "why" of applications on the job. In other cases the related courses, as in the case of technical writing, offer a vehicle of communication specifically designed to translate achievement into recognition of success. Individuals who teach related subjects must be able to strip an item of theoretical content and abstractions, and build a bridge from the unknown quantity to the topic essential for student achievement in a vocational sense. Lengthy proofs and deep abstractions have little interest for many students in vocational and technical programs. These students are action-oriented and hardware-minded. The most effective related subjects teachers not only know their subject matter in theory, but, they have the ability and the desire to translate this content into meaningful applications. Illustrations drawn from the shop or laboratory setting are useful in arousing and holding the interest of the students. Such illustrations can be readily obtained from those who teach in the shop or laboratory and it is imperative that the related subjects teacher use such illustrations. This means that the related subjects must be correlated with the shop or laboratory activity.

Some related subjects teachers have had actual work experience. This is not only helpful but highly desirable as it adds a dimension of understanding which improves communications between students and the teacher. Some states require varying degrees of work experience in a vocational field for those who seek a credential as a related teacher.

The subject matter competency of the related teacher must be considerably higher than the content which the person is teaching. Since there is a strong tendency for teachers to teach the way they were taught, it is a real challenge for these teachers to adapt the content to the needs of students and the time available in vocational and technical programs.

In some cases, where the related instruction takes place is also important. For example, in a senior high school serving many disadvantaged students in a building construction class, it was found most effective to have the related instructors teach the related material directly in the shop. Team teaching became the solution. The related mathematics instructor came into the shop to teach the math essential for applications to the trade. Likewise, the related English teacher carried on the instruction in English directly in the shop. The need might be writing a letter to a contractor whose housing project the students wanted to visit. This was done directly in the shop. Team teaching requires much cooperative planning. This situation demanded related teachers willing to work directly in the shop. It was an effective solution to a problem which previously had resulted in students rejecting the related instruction as not relevant.

The ability to teach is natural to some people, but most must learn how to teach just as they master other skills and knowledges. Teachers of related subjects must know how to teach if they are to be effective and the various techniques described elsewhere in this book must be mastered since a variety of approaches is often necessary. Related teachers must recognize the importance of preparation, planning, and organization of the content. But they must also be sufficiently flexible and adaptable to meet the changing needs of students and to provide the motivation needed for learning. How students learn is a vital part of the understandings of related teachers. Timing, relevance, and reinforcement are important in effective related instruction. The related teacher uses demonstrations, problem solving, and the conference method together with many other methods in "putting across" the subject. Likewise, the related teacher uses many visual and audiovisual aids and he must be familiar with these devices. The related instructor may teach large groups, small groups, or deal individually with students, so he must know how to operate effectively under varied conditions.

If the related instructor recognizes the applications of the content of his course to the goals of the students who are studying in the occupational program he will have an excellent opportunity for reaching and holding the interest of the students. He must always remember that he is not teaching mathematics, or science, or blueprint reading. He is, instead, teaching Jack or John or Jim how to solve a problem. He is teaching students. If the related instructor will approach his work in this way and with this philosophy he will be successful and will also achieve much greater satisfaction from his efforts.

What is the role of the related teacher in relation to that of the shop or laboratory teacher? If the related teacher will see himself as performing a vital service to the students and the teachers of the specialty he will more likely be in true perspective. He might think of himself in a position comparable to one in industry. In industry there are production departments and there are service departments. If the firm is in the business of manufacturing automobiles, some departments are producing parts; others are assembling parts; while others are testing the assembled units. These are in the direct line of production. Still other departments service the production functions, as, for instance, the personnel department, the accounting department, and the public relations department. They are not producing automobiles, but they are producing a vital service. The related instructor is more similar to the service department than the production department. One role is not necessarily better than another, but the roles are different. Each person must see his own role in true perspective to operate effectively and to provide the service to students that he is employed to provide.

Some related instructors seem humiliated to think that they are given the task of teaching vocational and technical students. This has sometimes been detected when a person who teaches, for instance, mathematics for college-bound students, is given the assignment of teaching a related class for vocational and technical students. It may well be that this individual is not competent to teach the related class, but it is a fact that some teachers have little empathy for vocational and technical students. It does not take long for students to detect this feeling. Such instructor attitudes seriously impede student learning and also make cooperation with the shop and laboratory instructors difficult. Individuals who lack the interest or ability to relate effectively to the students and the teachers with whom they should work cooperatively should be eliminated from teaching these courses.

Mathematics for Trade and Industrial and Technical Education

Mathematics is a tool subject. As a tool subject it performs two basic functions. *First,* it may be used as a method in teaching principles of science or technology. *Secondly,* it is necessary for calculations and computations essential for operations and jobs to be performed in the shop or laboratory.

If the instructors teaching the related mathematics have a background of employment in industry, or even a vocational orientation, it is quite easy to place mathematics in proper perspective.

Most technical studies and practically every trade require some proficiency with mathematics. But the type of mathematics and the depth of mathematical theory vary considerably from field to field. Often mathematics has been one of the major stumbling blocks for many of the individuals who enter programs in vocational and technical education. They have floundered on abstract theory in classes where instructors failed to indicate a real reason for learning mathematics or to indicate the importance of mathematics as a tool for accomplishing jobs that would be done in the shop or laboratory.

Teaching related mathematics necessitates identifying the level of knowledge possessed by the students as they enter class. This may be done by giving a pre-test. Recent public reaction against pre-tests, especially as used with minority groups, has reduced the number of schools using this approach. An alternative approach is to start at a comparatively low level with realistic problems and move rapidly through the content which does not present difficulties for the student. In many classes great variation in competency will exist between students. This may be solved in several ways such as:

1. Individualized instruction
2. Small group instruction
3. Using the more advanced students to aid the others
4. The buddy system

Individualized instruction in mathematics classes may consist of programmed learning. The armed forces has used this method for remedial instruction in mathematics. Another form of individualized instruction may use a well-written textbook and workbook as basic tools. The teacher works independently with students as problems appear.

Small group instruction is an attempt to better meet the needs of the students through homogeneous groupings or grouping students as nearly as possible according to background and rapidity of progress. Even though it may be possible to divide an average class into four or five approximately homogeneous groups, it is entirely possible that the groups will not remain homogeneous very long. Teachers who use this plan then work briefly with each group while the other students continue with their assignments. While not as satisfactory as individual or individualized instruction it does meet the needs of more individuals better than the blackboard illustration, problem-solving technique frequently employed by teachers of mathematics with large groups.

Since mathematics is a building-block type of subject it is necessary to proceed in a step-by-step process. Often the difficulty of students can be traced to instructors who in explaining solutions jump over steps, combine steps, or work the steps mentally and only write the partial answer on the blackboard. Even though mathematics requires a step-by-step solution, it does not follow that detailed proofs are required or that abstract theories need to be explained in detail. Keep in mind that students in vocational and technical education need to use mathematics as a tool to make a calculation which will permit continued progress towards the achievement of the operations or jobs assigned.

The faster students often can be very helpful in working with the slower students. This provides reinforcement for the fast students and aids the slow students at the same time. In this case it is also true that if you really want to learn a subject try to teach it to someone else.

The ability of students to communicate with each other and the closer status level of students in the same class indicate some of the advantages of this approach. This needs to be used in a limited manner and only if students are available who have the capacity and the desire to aid the slower students. Then, too, this must be carefully supervised by the teacher and kept within controllable limits.

The buddy system is a plan where a faster student may be paired with a slightly slower student on assigned study work. Again, caution must be exercised so that both individuals actively work the problems assigned. There is sometimes a tendency for one student to do all the work, resulting in very little learning for the other student. Close supervision on the part of the teacher is essential with this approach.

Often a combination of approaches is used by the effective teacher of related mathematics. One of the major problems is to motivate students and increase their interest in mathematics. With increased interest, it is common to find improved ability to perform. Problem solving requires much repetition and reinforcement. Student success leads to greater confidence. The wise teacher works to develop this kind of multiplier effect. The greatest contribution the teacher can make to the student is to increase the student's interest and desire to apply mathematics to problems arising out of the trade or technical field. Many do not feel that the traditional mathematics courses produce this effect. The traditional mathematics instructor often fails to organize his course to meet the needs of vocational education. The content of mathematics courses is specifically determined by the content of the shop and laboratory courses, which in turn can best be determined by analysis of the tasks actually performed on the payroll job. The selected topics need to be thoroughly taught but many topics can be omitted since they have little value for students in trade and technical programs.

Science for Applications

Just as the average vocational and technical student does not fit well into the traditional mathematics courses, he likewise does not belong in the science courses, as usually taught for college transfer students.

The base of technology is science. The average technician or tradesman needs the ability to apply certain scientific principles in his particular field. While science is often taught with heavy mathematics involvement, the applied sciences needed for applications in industry by the tradesman or technician can be taught in most cases without using the advanced mathematics.

First, the decision must be made as to what principles of science are important for the curriculums served. Again, this decision can best be made on the basis of the analysis of the occupation and the specific tasks required of tradesman and technician on the job. Advisory committees can provide invaluable assistance in making such identifications if a detailed study of industry has not been made. Once the decision has been reached as to the topics of science essential for the total program, the course of study can be constructed for the science course(s).

A problem may exist in finding suitable textbooks for the level of science or mathematics possessed by the students. If a suitable textbook is not available the teacher may need to build a series of information sheets and assignment sheets to aid the students as they study reference books and other available text materials. Publishers are more and more realizing the need for textbooks of technical science written at the level of the technician and tradesman. Unfortunately, these are often inadequate for most vocational fields.

Selected topics of physics and chemistry constitute the scientific emphasis for most of the fields in trade and technical education. In the health areas of technical education, biology and microbiology are essential. Again, selected topics can be delineated and taught without pursuing the topics from the same approach as that of the pre-medical courses.

The instructors of applied sciences need to relate the scientific principles to applications that occur in the industrial–technical world of work. With careful planning this can be done in such a way as to generate interest and motivate the students to further study on their own.

Individuals selected to teach courses in all applied sciences must work closely with the shop and laboratory instructors if maximum learning is to occur. Illustrations and problems can be extracted from shop and laboratory applications and used as the approach to the problems studied in the applied sciences classes. Stripping the application of the complex scientific theory and translating the application into terms, concepts, and uses within the grasp of the students involved is the major task of the instructor. In most cases this can be done without using mathematics of complexity beyond the students' comprehension.

The problem for most instructors is that they try to teach others the same way they were taught. Often, this involves the use of higher mathematics and complex scientific experiments.

It is highly desirable that all instructors of applied science courses have experience in industry at the level at which they are teaching. If this is not possible, individuals should be selected who are capable of working closely with the shop and laboratory instructors and who have the ability to translate science into meaningful experiences for the students in these curriculums.

Close correlation of the courses in mathematics and science with those in shop and laboratory is essential if the student is to benefit from the total program. Once the courses of study have been developed to achieve this objective, the timing of the course offering must be planned to integrate the related courses at the required time so that the content will be covered prior to or at the same time that the applications are being made in the shop and laboratory. Sequencing and scheduling of courses must reflect this need. During the first quarter or semester the time devoted to applied science may be much less than to shop or lab and mathematics. As the foundation is established in the shop or laboratory courses and the mathematics base is built, more emphasis can and should be given to the applied science. In other words, students are mainly interested in shop or laboratory courses in vocational education; therefore, use these to build interest and create holding power. Then draw from these experiences the needed emphasis in mathematics to permit continued growth and progress in the shop and laboratory. At the same time, relate the applied sciences to the shop or laboratory and employ the mathematics as a tool to produce solutions to needed problems. Close cooperation between the instructors and total integration of the three subject matter areas offer the best solution for maximum student achievement.

Communication for Students in Trade and Technical Education

Trade and technical students frequently are weak in communications—especially in written communications. Communication is essential in both business and industry. Ideas and concepts must be mobile. They must be moved from one individual to another as needed. Blocks to such movement must be removed in the school experience of the student if he is to work effectively in many industrial and service positions.

What kind of communication is important for students in trade and technical education? Again the answer can be found in the same way that answers are found to other curriculum needs—through analysis or through the use of advisory committees. Once the needs have been isolated then steps can be taken to build into the school program those experiences which are necessary. This will result in relevant education for the student and improve rather than retard his self-concept.

ORAL AND WRITTEN COMMUNICATIONS

The major problem with many efforts to teach communications lies in the failure to teach vital principles in terms of meaningful applications. Writing of a technical nature is essential for some technicians. Business letters and reports often need more attention for trade and technical students.

Oral communications in the form of brief reports and short talks have value for students in this field. Again the emphasis can easily be determined through a study of the trade or technology for which the program is intended.

Who should teach oral and written communications? Certainly the individual must have the ability to translate basic fundamentals of written and oral communications into uses meaningful for the student. Topics selected are more interesting if the student can write or speak about matters of interest to him such as activities of the shop or laboratory. Letters written may focus on needs arising within the scope of the specialty. Written reports may assume some of the features of technical reports, technical logs, and other similar types which technical and trade personnel may be expected to write. If the instructor is experienced in the trade or technical field he can use this background to good advantage in helping to identify applications-type content. If he is not familiar with the occupations, he needs to work closely with the shop and laboratory teachers to determine content which will be relevant for the students. The ability of the instructors to establish good relationships with the students is important in gaining their confidence and good will. Most vocational and technical students have been exposed to many years of oral and written English, but in many cases they have learned little as the content, emphasis, and approach failed to stimulate interest and were not connected with topics which seemed relevant to them. The author recalls the tremendous impact of an enthusiastic teacher of written and oral communications on students in a post-secondary technical institute. This teacher had several years of experience as a technical writer. He also had been an active member in Toastmasters' Clubs. His experiences, personality, and interest in the students generated a vibrant enthusiasm in a short while with students who approached his classes the first day with serious misgivings. Their "ho-hum" attitude disappeared. Much was accomplished in a few brief weeks. The instructor and the content of the subject made the difference for these students.

GRAPHIC COMMUNICATIONS

People who work in technical and trade occupations frequently need to make sketches and simple drawings. The ability to communicate through graphics is very helpful.

Some students have never had the opportunity to learn how to translate ideas into illustrations or sketches. Again the extent to which this should be included in the curriculum can be determined through analysis and through consultation with advisory committees.

Often some emphasis is given to sketching or drawing in connection with courses in blueprint reading if separate courses are not included in the curriculum.

Activities occurring in the shop or laboratory suggest many suitable items for teaching sketching or drawing to students.

Human Relations—Work Adjustment

It is a common practice in four-year college programs to require courses in humanities and social science. These courses are intended to help the individuals adjust better to other individuals, be more considerate of others, have greater appreciation of society, and be able to work more effectively both with the peer group and with superiors and subordinates. The same need frequently exists for those enrolled in vocational–technical curriculums.

The approach in vocational and technical schools usually is to offer a course or some units relative to human relations. The fundamentals of good human relations are discussed. Cases may be studied, situations may be simulated (role playing), and efforts may even be made to use sociodrama, or psychodrama. Sometimes committee work may be assigned with the focus on improved human relations.

The instructor may attempt to motivate students to improved relations within the organization (internal human relations) and try to prepare them to present a better image to the public at large. The latter is sometimes described as external human relations.

Can human relations be learned in classes? This is a question some people are asking. Many believe that human relations cannot be taught in classes whether these classes are in humanities and social science or in human relations. The best environment for learning human relations is in the laboratory of life. However, students need to acquire competency in human relations before employment.

Vocational and technical schools are fortunate in having shops and laboratories similar to those in which the students will work when they enter the world of work. These shops and laboratories provide excellent environments for applying the concepts of human relations acquired in classes designed to teach the subject. Again, close coordination between the shop and laboratory instructors and the instructor teaching the human relations class is essential. The shop or laboratory climate must reflect good internal human relations. The instructor must be aware of the opportunities and challenges for building in each of the students the attitudes and habits essential for adjusting to each other and to the teacher. As responsibility is assumed by students, cooperation is developed, and leadership functions are integrated with the operations and jobs of the shop or laboratory, students must be motivated to grow in human relations skills. The shop or laboratory is the best opportunity for mastery of the theory of human relations. But, such growth will occur most effectively in a climate planned to achieve this goal.

Opportunities also arise in connection with many of the activities of the shop or laboratory to generate effective external relations. In some cases promotions, advertising, and public services are integrated into the functions performed in the shop or laboratory. All of these activities provide opportunities for growth in external human relations.

The alert shop or laboratory instructor can do much to prepare the student for a smooth and easy transition in working with other people. This is an area of concern which deserves a high priority as a concomitant activity of all instructors in all formal and informal situations within the school environment.

Individual attention may need to be given to some individuals who have a serious problem. Referral to specialists may be a necessary solution. Most students will acquire adequate skill and knowledge in human relations through the informal and the planned inclusion of courses or units of courses. However, every teacher should be alert to needs and problems in this field. An individual may be an efficient and competent worker in his specialty, but he will have trouble holding his job if he cannot get along with people. Likewise, he will reduce his opportunity for advancement on many jobs if he cannot work effectively with people. Vocational and technical students and teachers must be "people" oriented.

The related teachers, as well as the shop and laboratory teachers, have an important role in integrating good work adjustment attitudes and habits with the content of the courses which they teach. It has frequently been said that more people lose their jobs because of their inability to get along with other people than because of lack of technical skill or knowledge.

Summary

Teachers of related subjects teach such subjects as related mathematics, science, communications, and work adjustment (human relations) subjects or units. Sometimes the shop or laboratory instructor teaches the units in related subjects, but more frequently this is done by special teachers.

Teachers of related subjects need to identify content in the same manner as shop or laboratory teachers—by analysis and/or through advisory committees.

Teachers of related subjects must:

1. Be competent in subject matter.
2. Possess the skill and knowledge to teach the subject to others.
3. Be genuinely interested in vocational and technical students.
4. Work closely and cooperatively with both shop and laboratory teachers.

The biggest challenge of related subjects teachers is to make the subject relevant and interesting for the students. To do this, each topic must be carefully selected, related to the operations and jobs of the shop or laboratory, and scaled down to the level of the students. Timing is important. Topics in mathematics and science need to be taught slightly before or at the same time as applications are first made of these concepts and principles in the shop or laboratory.

Work adjustment principles can be taught in class, but the applications of these principles must occur in the shop or laboratory. A unified approach with close cooperation between the related teacher and the shop or laboratory teacher is essential if the desired impact is to be made on the student relative to work adjustment and human relations.

QUESTIONS AND ACTIVITIES

1. How should the content of the course of study for each class of related instruction be determined?
2. What should be the requirements for the credential of related subjects teachers? Should work experience in a trade or technical area be required?
3. Should related subjects teachers teach mathematics for mathematics sake? Explain.
4. Do you believe that physics or other sciences can be taught with low-level mathematics? Discuss.
5. What procedures are essential for integration of the work of the related classes with that of the shop or laboratory classes?
6. To what extent is team teaching adaptable to related subjects classes?
7. Discuss the importance of communications for students of trade and technical curriculums.
8. Is sketching a form of communications? Mathematics? Explain.

9. To what extent can human relations be taught in class? Discuss.
10. Why do students in trade and technical classes have a better opportunity to master effective human relations than in many of the academic courses?
11. Select three problem situations which may arise in the related classroom and explain how these difficulties may be worked out. Select one of these situations and through role playing act out the solution to the problem in class.
12. What is the role of the related subjects teacher in the total education of the trade and technical students?

Chapter 4 IDENTIFYING VALID INSTRUCTIONAL CONTENT

Curriculum construction based on employment needs is the essence of effective payroll education for the youth and adult in today's world. The climate of society and of recent legislation has highlighted the need for valid instructional content in both high school and post-high school instructional programs.

The real thrust of building curriculum for vocational instruction is found in analysis of payroll jobs and occupations. The translation of the employer's needs for personnel competencies today and tomorrow can only be validly reflected in the instructional program after identifying what employees do on the job, and how the employer anticipates the jobs will be changed in the future.

Job analysis has been used for many years. A long time ago, Comenius, a Czech, believed that youth would discover their special aptitudes if they were given instruction in handwork along with academic subjects. In the latter part of the eighteenth century, Pestalozzi established an educational system built around farming, spinning, and weaving. Otto Saloman of Sweden advocated a plan which focused attention on the analysis of operations and the educational experiences that resulted from such analysis. In the United States, Charles R. Allen stimulated much interest in the process of job analysis for curriculum construction in vocational education.

The early efforts in analysis can be traced to the adoption of the concept of teaching vocational education through jobs, questions, problems, and guided discussions. This concept is still a valid and common basis for much of the instruction in vocational and technical education.

The increasingly complex nature of today's technological society mandates more refined methods in arriving at that content which is essential for an educational program designed to prepare students for earning a living as well as for life itself. Many of the problems of unemployment and underutilization of manpower are the direct result of inadequate analysis of occupations and poorly constructed educational programs that prepare students for nothing but unemployment and relief subsidies at the expense of others.

The state and nation can no longer afford unproductive educational experiences for large numbers of students. Reinforcement of traditional school experiences in the world of work is the result of such identified needs. More realistic information must flow from the "user of the school product" to the "producer of the job applicant" if the needs of both the employee and the employer are to be met. Educational planning and structures for learning need to define essential, sequential relationships and provide for acquisition of capabilities in relevant content. This can all be done through occupational analysis.

An effective technical training program has five basic requirements as delineated in a paper presented to the National Society of Programmed Instruction. These are:

1. Job specification
2. Translation of job specification into training objectives
3. Measurement of the individual's aptitude for training
4. Development of a method of achieving the training objective
5. Performance control and evaluation (8, p. 1)

It has been established by another writer that underlying all the professional, skilled, and technical occupations was a set of behaviors which could be described and taught. (6, pp. 209–14)

Many definitions of job analysis have been formulated. The Department of Labor's *Training and Reference Manual for Job Analysis* describes job analysis as the process of identifying by observation, interview, and study of the technical and environmental facts of a specific job and reporting the significant workers' activities and requirements. (9) Another source characterizes job analysis as the process of studying the operations, duties, and organizational relationships of jobs to obtain data for writing job descriptions and job specifications. (5, p. 42) From the point of view of program planning and curriculum building, analysis is the identification, breakdown, and separation of competencies utilized in the performance of the task, operation, job, or occupation. Most readily identified are the skills employed. Essential knowledges are less discernible. Employable habits and desirable attitudes are even more elusive.

Sources

Information about tasks, operations, jobs, and occupations may be classified according to the sources from which the information was secured. Sources are either primary or secondary. Information obtained from primary or original sources is obtained directly from those actually involved. Mail surveys, interviews, and personal participation of the investigator are considered primary sources. Secondary sources reflect the recorded concepts of others rather than the observations of individuals directly involved in the job or occupation. Examples of some secondary sources are textbooks, manuals, and articles in periodicals.

Various authors have discussed the sources of materials available for the development of curriculum. One of these, Cay, mentions individuals, subjects, and the total society as three main sources. (1, pp. 7–8) While these sources give a broad setting for curriculum development, it is helpful to sharpen the focus. The practical sources for information about a curriculum for vocational and technical education are related to the job in industry. Information about the job can be gained by personal experience, observations of individuals performing the tasks, mail surveys, interviews with workers and their supervisors, and from members of advisory committees. Books, pamphlets, periodicals and other secondary sources also provide helpful material.

ORIGINAL SOURCES

The occupational or job analyst (just as the historian) who uses original sources is more likely to end up with more accurate information. Individuals who do not check the accuracy of the information used frequently pass on errors made by previous writers.

Utilizing original sources necessitates identifying the source, knowing what to observe, recording the information, and applying appropriate relative values to that which has been observed. For most effective results, analysts who are not familiar with the payroll job or occupation need to be oriented prior to assuming the task. Consistency in observation, questioning, and recording of observations is significant if the results are to be reliable.

Observed Activities. An excellent illustration of the procedure for job analysis of the observed activities has been developed by the Bureau of Employment Security, United States Department of Labor. (2) Job analysis is defined in terms of what the worker does, why he does it, how he does it, and the skills involved in the doing. Explanation of the key points and important considerations given in the guide follow.

WHAT THE WORKER DOES

What the worker does involves the physical and mental responses that are made to the work situation. Physically the worker may transport materials, cut, bend, grind, put together, make ready, set up, tear down, insert, regulate, clean, finish, or otherwise change the position, shape, or condition of the work by the expenditure of physical effort. Mentally the worker may plan, compute, judge, direct, or otherwise govern the expenditure of his own or others' physical effort by a corresponding exercise of mental effort. In a given job a worker may expend any combination of physical and mental effort required by the task.

In determining *What* the worker does, the analyst must establish the complete scope of the job and consider all of the physical and mental activities involved.

Most jobs involve more than one task and each task may involve different physical and mental activities. The analyst must determine the tasks and report and describe them in direct terms so as to present a clear and coherent work picture. This may be done by describing the tasks chronologically or by grouping them according to their nature, depending on which creates the clearest presentation.

Considerations:

What tasks have been observed in the job?

Are there additional tasks which have not been observed? Are these additional tasks customary for all workers on the job?

Are the tasks included for this job performed by all workers designated by the job title?

What is the frequency with which the tasks are performed?

What is the relative difficulty of each task as compared with the rest of the tasks of the job?

Has the data obtained by observation been verified with the proper authority?

HOW HE DOES IT

How the work is done concerns the methods used by the worker to accomplish his tasks. Physically this involves the use of machinery and tools, measuring instruments and devices, and other equipment, the following of procedures and routines, and the movements of the worker himself. Mentally the methods lie chiefly in the "know how" that must be applied to the tasks. This may involve the use of calculations, formulas, the application of judgment or decision, or selection and transmittal of thought. The worker may use a single method in the accomplishment of a task or he may have at his command several alternate methods, any of which may be used with equal success.

Considerations:

What tools, materials, and equipment have been used to accomplish all the tasks of the job?
Are there other tools, materials, or equipment which have not been observed? If so, how do they work?
What methods or processes have been used to accomplish the tasks of the job?
Are there other methods or processes in the plant by which this same work can be done? If so, what are they?

WHY HE DOES IT

Why a worker performs his job is the purpose of the job itself and is indicative of the relationships among the tasks that comprise the total job.

The *Why* outlines the scope of the job and justifies the *What* and *How* of the work performed. The overall purpose, of course, is the sum total of the purposes of all tasks. Utmost care is necessary in ascertaining and recording the reason why each task is performed, both to clarify the overall purpose of the job in the reader's mind and to show task relationships as the job progresses to the completion of a work cycle.

The purpose may be the conversion of material from one form to another, the maintenance of conditions under which other jobs can be performed, the catching or preventing of errors, the development of new methods or the improvement of existing methods, and so forth. Failure to explain this purpose will leave the impression that the job has not been reported completely, and tends to ambiguity in the description of the job.

The purpose or *Why* of the job is the first thing the analyst must ascertain in order to orient himself for the subsequent analysis.

Considerations:

Why is the job done? What is its overall purpose?
Why is each task done? What is the purpose of each?
What is the relationship of each task to other tasks and to the total job?

THE SKILL INVOLVED

This part of the Job Analysis Formula brings out important information necessary to supplement the *What, How,* and *Why* and to express the degree of difficulty of the work tasks involved in the job.

It consists of a listing and an explanation of the basic factors which must be considered in analyzing any job. These elements bring out the manual skills, knowledges, abilities, and other characteristics required of a worker by his job, regardless of whether that job is manual, craft, professional, clerical, or other type. It may be considered a guide list to aid the analyst in obtaining and recording all the information necessary to discriminate between jobs and to establish the degree of difficulty of any job.

RESPONSIBILITY

Considerations:

Supervisory Responsibility

How closely are subordinates supervised?
How many workers are supervised and what are their duties?
Is the supervision direct or indirect?
What supervision is received by the worker?

Nonsupervisory Responsibility

What tools, equipment, materials, or product could be lost by work failure? What would be the value lost?
What time loss would be caused? What would be its value?
What is the likelihood of loss and what can the worker do to prevent it?
What injuries can occur as a result of work failure?
If others can be injured what can they do to protect themselves or are they entirely dependent on this worker?
What safety devices or checks exist?
Is worker required to cooperate with other workers? How and to what extent?
Is worker responsible for contacts which include outsiders? In what manner?

JOB TRAINING

The practical knowledge of equipment, materials, working procedures, techniques, and processes which the worker must possess to handle the job successfully.

Measured by:
Amount and complexity of practical knowledge which the worker must possess, whether this knowledge is gained by actual on-the-job experience, prior training, or by both.

Limited by:
Nature of instructions received regarding work tasks.
Degree of supervision received.

TRAINING

This may be considered as the place where, the time in which, and the method by which required physical and mental skills are developed by the worker. It is exclusive of experience but is similar in that, while not a part of job achievement, it is indicative that the *skill involved* in the job is possessed by the worker in the required amount and kind. If it is the minimum acceptable for job placement, it is required. If it is over and above the minimum requirement but still contributes to job success, it is desirable. In either case, it is important to the placement officer either for recruitment or for guidance.

Measured by:

The kind of training required and the time spent in acquiring it.

Limited by:

The job knowledge and dexterity and accuracy which are acquired during the training.

The job knowledge and dexterity and accuracy which are not acquired but in which the worker must receive additional training to become proficient. (2, pp. 2–25)

Questionnaire and Interview Surveys. Information about a single payroll job, an entire occupation, or a particular industry can be secured by means of a survey. Two approaches commonly used are mail questionnaires and interview schedules. The questionnaire is a form distributed through the mail or filled out by the respondent under the supervision of the investigator, while the schedule is a form filled out by the investigator in the presence of the respondent. Most of the concepts and techniques of the questionnaire are equally applicable to the interview.

There are various types of instruments that may be used for either the questionnaire or interview schedule. The closed-form questionnaire is used when categorized data are desired. The informant chooses from a set of provided responses and merely checks or otherwise marks the response which most closely approximates or describes the situation. The open-form or free-response questionnaire is used for intensive studies of a limited number of cases. The respondent is asked to write a descriptive statement to answer each question posed. The checklist is a set of categorized statements for the respondent to check. The dichotomous questionnaire is a type of closed-form instrument designed to provide opposite alternatives such as "yes" or "no," "agree" or "disagree."

The primary advantage of the interview over the questionnaire is its greater flexibility. The investigator may ask for clarification of points which are not clear and observe the respondent for signs of evasiveness, noncooperation, or other irregularities. The limitation of the interview results from the influence the investigator may exert on the respondent.

One common form of interview is the structured interview. This is a rigidly controlled interview using an interview schedule and following a predetermined uniform plan. The unstructured interview, on the other hand, often takes the form of a conversation with very few attempts to direct the line of discussion.

Structuring of the instrument is an important step in the planning of a survey. As a research tool it should be developed in conformance with good research technique. Another significant step is the determination of the group to be surveyed. Often a sample of the total population is chosen in accordance with the practice of the use of random sampling. If a desire exists to generalize to the total population, a plan of randomization is mandatory. The size of the sample can be determined statistically.

Three illustrations of forms used for collecting data on payroll jobs for the purpose of building curriculum are reproduced in this chapter. (See Exhibits 4–1, 4–2, and 4–3.) (Text continued on p. 67)

EXHIBIT 4–1

Forms Used for Collecting Job Data

Three forms have been adopted for collecting data to be used in selecting jobs to be included in the training curricula and to describe those jobs. Examples of these forms appear in the order described here. The examples shown are only partially complete.

1. Form 1 is designed to be used in listing jobs identified as being in the area under consideration by each of the coordinators. Provision is made for simply checking whether or not each job listed should be selected for more detailed description according to criteria supplied the coordinator and described in the document in Appendix A [of Morrison's report]. The last column, headed "Reason," provides space for indicating by code numbers the reasons for rejecting a job if it has been rejected.
2. Form 2 is designed to provide information needed to further screen and reduce the number of jobs to be included for complete description and inclusion in the training program. The form as it appears here is a revision of the form suggested in the document in Appendix A. Specifications of the information to be recorded on the form also appear in that document.
3. Form 3 provides space for listing tasks associated with each of the jobs described on Form 2. Provision is made for classifying tasks into categories suggested in the document mentioned above. The third and final column will be used in analysis to be conducted in subsequent phases of the project.

Source: Morrison, Edward J., *Development and Evaluation of an Experimental Curriculum for the New Quincy (Mass.) Vocational Schools.* Pittsburgh: American Institute for Research, June 30, 1965, Appendix 13, pp. 1–3.

EXHIBIT 4–1 continued

FORM 1

Electro-electronics (Electronics)

	1. JOB NAME	2. SELECT		3. REASON
		Yes	No	
1.	Electronic Technician			
2.	Wireman	✓		
3.	Senior Test Technician			
4.	Test Technician			
5.	Electro-Mechanic Assembler			
6.	Electrical Wireman			
7.	Apprentice Electronics Technician			
8.	Repair Technician			
9.	Wireman Assembler			
10.	Panel Assembler			
11.	Construction Technician			
12.	Environmental Technician			
13.	T.V. Serviceman			
14.	Senior Test Equipment Technician			
15.	Appliance Technician			
16.	Special Equipment Wireman			
17.	Field Technician			
18.	Electronic Assembler			
19.	Electronic Circuit Technician			
20.	Meter Technician			
21.	Microwave Technician			
22.	Communication Technician			

EXHIBIT 4–1 continued

FORM 2

Accept Reject

1. JOB FAMILY Electro-electronics	2. COORDINATOR(S) D. R. Kaupp	3. DATE 5/20/65
4. SUBFAMILY Electronics	6. LOWER LEVEL JOBS None	
5. JOB NAME Wireman 5a. EQUIVALENT NAMES May be combined with Assembler and/or Solderer		

7. PRE-EMPLOYMENT TRAINING TIME (Beyond 8th Grade)

ACADEMIC	VOC-TECH TRNG.	MILITARY EXP.	INDUSTRIAL EXP.
None	6 months	None	1 month

8. EMPLOYMENT QUALIFICATION REQUIREMENTS

PHYSICAL	LEGAL	TOTAL TRNG.	OTHER
Normal	None	7 months	

9. TRAINING FACILITIES REQUIRED Government, NASA, and a variety of commercial specifications as well as standard tools and equipment of the trade.

10. INDUSTRIES USERS All electronic manufacturing companies.

11. KNOWN CHANGES IN JOB REQUIREMENTS Increasing emphasis on miniaturization and solid state circuitry.

12. EMPLOYMENT OUTLOOK Local	National
Always in demand as this job function is fundamental to all production of electronic equipment. Demand expected to rise continually through next five years (*Occupational Outlook Handbook,* 1963, p. 587).	Same as local

EXHIBIT 4–1 continued

FORM 3

Electro-electronics (Electronics)

	1. TASK NAME (Wireman)	2. TYPE	3. CODE
	All tasks of the Solderer plus the following:		
1.	Stripping Wires & Cables		
2.	Uses Crimping Tool		
3.	Cables		
4.	Makes Wiring Harnesses		
5.	Makes Wrap Connections		
6.	Makes Screw Connections		
7.	Pulls Cables & Wires		
8.	Mounts Components		
9.	Reads Schematics Layouts		
10.	Reads Color Codes		
11.	Makes Continuity Checks		
12.	Selects Wires & Cables		
13.	Measures Wire Lengths		
14.	Installs Electrical Hardware		
15.	Insulates Wires		
16.	Cuts Wires		
17.	Reads Run Sheets, Specifications Sheets		

EXHIBIT 4–2

Instructional Analysis Questionnaire of the Automotive Mechanics Occupation

DEFINITION

(Remember this definition covers the duties of a fully qualified Automotive Mechanic and may contain more than is required at the entry level.)

AUTOMOTIVE MECHANIC (auto. ser.) 620.281. Repairs and overhauls automobiles, buses, trucks, and other automotive vehicles: Examines vehicle and discusses with customer or Automobile-Repair-Service Salesman; Automobile Tester; or Bus Inspector nature and extent of damage or malfunction. Plans work procedure, using charts, technical manuals, and experience. Raises vehicle, using hydraulic jack or hoist, to gain access to mechanical units bolted to underside of vehicle. Removes unit, such as engine, transmission, or differential, using wrenches and hoist. Disassembles unit and inspects parts for wear, using micrometers, calipers, and thickness gages. Repairs or replaces parts, such as pistons, rods, gears, valves, and bearings, using mechanic's handtools. Overhauls or replaces carburetors, blowers, generators, distributors, starters, and pumps. Rebuilds parts, such as crankshafts and cylinder blocks, using lathes, shapers, drill presses, and welding equipment. Rewires ignition system, lights, and instrument panel. Relines and adjusts brakes, aligns front end, repairs or replaces shock absorbers, and solders leaks in radiator. Mends damaged body and fenders by hammering out or filling in dents and welding broken parts. Replaces and adjusts headlights, and installs and repairs accessories, such as radio, heaters, mirrors, and windshield wipers.

Manipulative Skill Required Code

1. Speed required but very little skill required.
2. Reasonable time limit and a reasonable degree of skill required.
3. Reasonable time limit, but a high degree of skill required.
4. Speed and a high degree of skill required.

Technical Knowledge Required Code

1. Ability to do minor repair work following oral instructions.
2. Ability to read and interpret the manuals and charts, and to work under supervision.
3. Ability to work independently, using manuals and charts.
4. Ability to diagnose and troubleshoot.

DIRECTIONS

The purpose of this questionnaire is to determine the basic skills, knowledges, and other qualifications that are required of a worker to enter the occupation for the first time as an Automotive Mechanic. The following pages contain

Source: Instructional Analysis of the Automotive Mechanic Occupation. Tallahassee: Division of Vocational, Technical, and Adult Education, State Department of Education, May, 1967, pp. 1–3.

EXHIBIT 4–2 continued

an analysis of the Automotive Mechanic Occupation. The occupation is broken down to the basic operations and organized into blocks, units, and operations. Examine carefully each block, unit, and operation.

1. Behind each operation draw a circle around either the word "Yes" or "No," in the column headed REQUIRED FOR EMPLOYMENT, to indicate the operation is or is not required of the entry worker.

If "Yes" is circled in the column headed REQUIRED FOR EMPLOYMENT:

2. Indicate the amount of manipulative skill required by circling the appropriate number in the column headed MANIPULATIVE SKILL REQUIRED (refer to the code given previously).
3. Indicate the amount of technical knowledge required to perform the operation by circling the appropriate number in the column headed TECHNICAL KNOWLEDGE REQUIRED (refer to the code given previously).

Space has been provided at the end of each unit for other operations that need to be added to the list. If operations are added, indicate the amount of Manipulative Skill Required and the amount of Technical Knowledge Required of the entry worker.

BLOCK I. ENGINE

UNIT	OPERATIONS	REQUIRED FOR EMPLOYMENT	MANIPULATIVE SKILLS REQUIRED	TECHNICAL KNOWLEDGE REQUIRED
A. Head assembly	1. Replace head gaskets	Yes No	1 2 3 4	1 2 3 4
	2. Grind valves and seats	Yes No	1 2 3 4	1 2 3 4
	3. Repair valve train (except camshaft)	Yes No	1 2 3 4	1 2 3 4
	4. Diagnose valve train and head malfunctions	Yes No	1 2 3 4	1 2 3 4
B. Block assembly	1. Fit and install pistons, pins, and rings	Yes No	1 2 3 4	1 2 3 4
	2. Adjust or repair the crankshaft and connecting rod assembly	Yes No	1 2 3 4	1 2 3 4
	3. Replace timing chain and gears	Yes No	1 2 3 4	1 2 3 4
	4. Replace camshaft and/or camshaft bushings	Yes No	1 2 3 4	1 2 3 4

EXHIBIT 4–2 continued

BLOCK I. ENGINE (Continued)

UNIT	OPERATIONS	REQUIRED FOR EMPLOYMENT	MANIPULATIVE SKILL REQUIRED	TECHNICAL KNOWLEDGE REQUIRED
B. Block assembly	5. Replace engine and/or transmission supports	Yes No	1 2 3 4	1 2 3 4
	6. Service the engine breathing system	Yes No	1 2 3 4	1 2 3 4
	7. Replace oil filter	Yes No	1 2 3 4	1 2 3 4
	8. Diagnose malfunctions in the block assembly for each of the above	Yes No	1 2 3 4	1 2 3 4

BLOCK II. ELECTRICAL SYSTEM

UNIT	OPERATIONS	REQUIRED FOR EMPLOYMENT	MANIPULATIVE SKILL REQUIRED	TECHNICAL KNOWLEDGE REQUIRED
A. Cranking system	1. Service the battery	Yes No	1 2 3 4	1 2 3 4
	2. Repair cables, wiring, switches, relays, and solenoids	Yes No	1 2 3 4	1 2 3 4
	3. Repair the cranking motor and/or solenoid	Yes No	1 2 3 4	1 2 3 4
	4. Analyze malfunctions in the cranking system	Yes No	1 2 3 4	1 2 3 4
B. Charging system	1. Service the generator or alternator	Yes No	1 2 3 4	1 2 3 4
	2. Repair wiring	Yes No	1 2 3 4	1 2 3 4
	3. Make necessary adjustments to the regulator	Yes No	1 2 3 4	1 2 3 4
	4. Analyze malfunctions in the charging system	Yes No	1 2 3 4	1 2 3 4
C. Ignition system	1. Inspect and repair the switch, resistor, wiring, coil, points and condenser of the primary circuit	Yes No	1 2 3 4	1 2 3 4
	2. Inspect secondary circuit leads distributor, cap and/or rotor	Yes No	1 2 3 4	1 2 3 4
	3. Clean and test spark plugs	Yes No	1 2 3 4	1 2 3 4

EXHIBIT 4–2 continued

BLOCK II. ELECTRICAL SYSTEM (Continued)

UNIT	OPERATIONS	RE-QUIRED FOR EMPLOY-MENT	MANIPU-LATIVE SKILL RE-QUIRED	TECH-NICAL KNOWL-EDGE RE-QUIRED
C. Ignition system	4. Test ignition timing and advance units on a test stand	Yes No	1 2 3 4	1 2 3 4
	5. Set the timing of the ignition to the engine	Yes No	1 2 3 4	1 2 3 4
	6. Determine the reasons for weak or no spark at the plug	Yes No	1 2 3 4	1 2 3 4

EXHIBIT 4–3

INSTRUCTIONAL ANALYSIS QUESTIONNAIRE OF THE COSMETOLOGY OCCUPATION

DEFINITION

(Remember this definition covers the duties of a cosmetologist and may contain more than is required at the entry level)

COSMETOLOGIST (per. ser.) 332.271. Beautician; beauty culturist, beauty operator; cosmetician. Provides beauty services for customers: Suggests coiffure according to physical features of patron. Styles hair by cutting, trimming, and tapering, using clippers, scissors, and razors. Shampoos hair and scalp with water, liquid soap, dry powder, or egg, and rinses hair with vinegar, water, lemon, or prepared rinses. Applies water or waving solution to hair and winds hair around rollers, or pin curls and finger-waves hair. Applies bleach, dye, or tint, using hands or cotton pads, to color customer's hair, first applying solution to portion of customer's skin to determine if customer is allergic to solution. Suggests cosmetics for conditions, such as dry or oily skin. Applies lotions and creams to customer's face and neck to soften skin and lubricate tissues. Massages scalp and gives other hair and scalp-conditioning treatments for hygienic or remedial purposes (SCALP-TREATMENT OPERATOR). Performs other beauty services, such as massaging face or neck, shaping and coloring eyebrows or eyelashes, removing un-

Source: Instructional Analysis of the Cosmetology Occupation. Tallahassee: Division of Vocational, Technical and Adult Education, State Department of Education, September, 1968, pp. 1, 3–5.

EXHIBIT 4–3 continued

wanted hair, applying solutions that straighten hair or retain curls or waves in hair, and waving or curling hair. Cleans, shapes, and polishes fingernails and toenails (MANICURIST). May be designated according to beauty service provided as FACIAL OPERATOR: FINGER WAVER: HAIR TINTER, MARCELLER: PERMANENT WAVER: SHAMPOOER.

Manipulative Skill Required Code	Technical Knowledge Required Code
1. Speed required, but a high degree of skill not required.	1. Low degree of technical knowledge required.
2. Reasonable time limit, but a high degree of skill required.	2. Medium degree of technical knowledge required.
3. Speed and a high degree of skill required.	3. High degree of technical knowledge required.

DIRECTIONS

The purpose of this questionnaire is to determine the basic skills, knowledge, and other qualifications that are required of a worker to enter the occupation for the first time as a Cosmetologist. The following pages contain an analysis of the Cosmetologist Occupation. The occupation is broken down to the basic operations and organized into blocks, units, and operations. Examine carefully each block, unit, and operation.

1. Behind each operation draw a circle around either the word "Yes" or "No," in the column headed REQUIRED FOR EMPLOYMENT, to indicate the operation is or is not required of the entry worker.

If "Yes" is circled in the column headed REQUIRED FOR EMPLOYMENT:

2. Indicate the amount of manipulative skill required by circling the appropriate number in the column headed MANIPULATIVE SKILL REQUIRED (refer to the code given previously).
3. Indicate the amount of technical knowledge required to perform the operation by circling the appropriate number in the column headed TECHNICAL KNOWLEDGE REQUIRED (refer to the code given previously).

Space has been provided at the end of each unit for other operations that need to be added to this list. If operations are added, indicate the amount of Manipulative Skill Required and the amount of Technical Knowledge Required of the entry worker.

EXHIBIT 4–3 continued

BLOCK I PERMANENT WAVING

UNIT	OPERATIONS	REQUIRED FOR EMPLOYMENT		MANIPULATIVE SKILLS REQUIRED	TECHNICAL KNOWLEDGE REQUIRED
A. Cold Waving	1. Part	Yes	No	1 2 3	1 2 3
	2. Section	Yes	No	1 2 3	1 2 3
	3. Roll with paper	Yes	No	1 2 3	1 2 3
	4. Roll with lamb's wool	Yes	No	1 2 3	1 2 3
	5. Make test curls	Yes	No	1 2 3	1 2 3
	6. Selection of types of solution, and tools, and equipment	Yes	No	1 2 3	1 2 3
	7. Observing safety measures	Yes	No	1 2 3	1 2 3
B. Cold Waving Tinted or Bleached Hair	1. Use special blocking for damaged hair	Yes	No	1 2 3	1 2 3
	2. Roll hair	Yes	No	1 2 3	1 2 3
	3. Time the solution action	Yes	No	1 2 3	1 2 3
	4. Make record of cold wave	Yes	No	1 2 3	1 2 3
	5. Make microscopic examination of hair	Yes	No	1 2 3	1 2 3
	6. Observe safety precautions	Yes	No	1 2 3	1 2 3
C. Heat Waving	1. Able to follow heat waving instructions	Yes	No	1 2 3	1 2 3
	2. Block and roll hair	Yes	No	1 2 3	1 2 3
	3. Differentiate difference in heat and cold wave rolling	Yes	No	1 2 3	1 2 3
	4. Arrange machine or machineless heat waving	Yes	No	1 2 3	1 2 3
	5. Make test curls in heat wave	Yes	No	1 2 3	1 2 3
D. Permanent Waving for Pressed Hair	1. Marcel waves	Yes	No	1 2 3	1 2 3
	2. Croquignole waves	Yes	No	1 2 3	1 2 3
	3. Techniques of cold waving	Yes	No	1 2 3	1 2 3
	4. Work precautions	Yes	No	1 2 3	1 2 3
	5. Trade names used	Yes	No	1 2 3	1 2 3

EXHIBIT 4–3 continued

BLOCK I PERMANENT WAVING (Continued)

ADDITIONAL OPERATIONS AND COMMENTS:

BLOCK II HAIR CUTTING AND SHAPING

UNIT	OPERATIONS	REQUIRED FOR EMPLOYMENT	MANIPULATIVE SKILLS REQUIRED	TECHNICAL KNOWLEDGE REQUIRED
A. Basic Hair Shaping	1. Use correct type implements for hair shaping	Yes No	1 2 3	1 2 3
	2. Find natural growth of hair	Yes No	1 2 3	1 2 3
	3. Part for a basic hair cut	Yes No	1 2 3	1 2 3
	4. Operate scissors and comb	Yes No	1 2 3	1 2 3
	5. Use thinning shears, and recognize where and when to thin	Yes No	1 2 3	1 2 3
	6. Use razor, cutting and thinning the hair simultaneously	Yes No	1 2 3	1 2 3
	7. Thin with regular scissors	Yes No	1 2 3	1 2 3
	8. Know procedure for a basic hair cut	Yes No	1 2 3	1 2 3
	9. Care for tools	Yes No	1 2 3	1 2 3
B. Style Cutting	1. Teenager	Yes No	1 2 3	1 2 3
	2. Career girl	Yes No	1 2 3	1 2 3
	3. Fitted neckline	Yes No	1 2 3	1 2 3
	4. Style of the month	Yes No	1 2 3	1 2 3

ADDITIONAL OPERATIONS AND COMMENTS:

EXHIBIT 4–3 continued

BLOCK III SCALP TREATMENT

UNIT	OPERATIONS	REQUIRED FOR EMPLOYMENT	MANIPULATIVE SKILLS REQUIRED	TECHNICAL KNOWLEDGE REQUIRED
A. Scalp Treatments for Dry Scalp	1. Brush hair	Yes No	1 2 3	1 2 3
	2. Manipulation of scalp	Yes No	1 2 3	1 2 3
	3. Apply oil	Yes No	1 2 3	1 2 3
	4. Use safety precautions with heat cap	Yes No	1 2 3	1 2 3
B. Dandruff Treatment	1. Apply dandruff lotion or ointment to the scalp	Yes No	1 2 3	1 2 3
	2. Use the steamer	Yes No	1 2 3	1 2 3

Consulting Committees. Another method frequently used to develop valid curriculum is the consulting committee, frequently referred to as the advisory committee. Review of the literature reveals many illustrations of teachers working with advisory committees in the development of a guide for the planning of a curriculum. The common plan consists of having the committee review the plans and procedures used in the development of the curriculum and courses of study. Later the committee is given the opportunity to study and make comments for improvement. Much of the effectiveness of this procedure depends upon the personnel of the consulting committee and the way in which the teachers have involved the committee in the developmental process.

A study by Richmond Union High School District and Polytechnical College illustrates the use of very active study committees. (10) The Richmond Plan was the study of a program designed to meet the needs of the average high school student by establishing a pre-technical curriculum in two high schools and in Cogswell Polytechnical College.

Consulting committees are frequently employed to authenticate the curriculum of State Departments of Vocational Education and of some of the curriculum laboratories.

The role of the advisory committee in occupational education is discussed in several publications including a very effective one by the American Association of Junior Colleges. (11)

Personal Participation. Another method of acquiring information of changes in employment which should be reflected in the curriculum and the courses of study is to secure employment in the field. This provides opportunities for personal updating and an informal opportunity to acquire firsthand information of the changing emphasis in industry, business, and agriculture. Each individual usually does not have enough time to sample broadly employment requirements and job specifications. However, the personal background of the curriculum developer is important and must not be ignored.

SECONDARY SOURCES

Often, curriculums are developed on the strength of research or analysis performed by others. Some of these secondary sources are the products of valid analysis and careful occupational research, but many are not.

Study of textbooks and reference books provides insight into the efforts of others even though these may not be valid from the point of view of occupational analysis.

Handbooks, manuals, and other reference works developed by manufacturers also provide resource material which should be considered in assessing resources for curriculum development.

Articles in periodicals, government pamphlets, and releases of professional societies likewise should be reviewed before the final decision relative to curriculum is made.

It is wise to recognize that many publications which do not indicate analysis of payroll jobs or occupations most likely have been developed without such validation.

Summary

The value of the curriculum to the student as preparation for entry into an occupation is dependent upon the accuracy of the sources of information used.

Two sources of information about occupations are primary and secondary sources. Primary source information is the result of the investigation of the researcher, while secondary sources reflect the concepts of other writers.

Primary sources of occupational information are:

1. On-site study
2. Questionnaire (by mail)
3. Consulting committee

The study of occupations needs to produce information as to:

1. What the worker does
2. How he does it

3. Why he does it
4. The skill involved
5. Responsibility
6. Job knowledge
7. Training

Mail questionnaires may be open-ended or closed-form instruments. Random sampling and size of sample should be applied in conformity with good research methodology if generalization to the total population is to be made.

Mail questionnaires need to provide information as to:

1. Whether or not the operation is an essential part of the work of the employee in the occupation.
2. What degree of manipulative skill and technical knowledge is required to perform the operation.

Frequent sources of occupational information are consulting committees and personal participation.

QUESTIONS AND ACTIVITIES

1. Why should considerable effort be given to validation of curriculum content?
2. What is meant by a performance based curriculum?
3. Trace the history of occupational analysis.
4. Discuss the use of analysis in industry. In business. In the armed forces. In research.
5. What are the requirements for an effective training program?
6. Define job analysis.
7. What are the possible referents for development of curriculums?
8. Delineate between primary and secondary sources of information about occupations.
9. In making an analysis of an occupation explain the importance of:
   a. What
   b. How
   c. Why
10. In addition to needed skill what other information should be obtained by the job analyst?
11. What are the advantages of the open-ended questionnaire? The closed-form type of instrument?

12. Why should random sampling be employed in relation to the mail questionnaire?
13. Is size of sample used with a questionnaire study important? Explain.
14. Why is the amount of skill and knowledge used an important part of a mail questionnaire study of occupations?
15. A curriculum is being developed. A new consulting committee is being formed. What criteria should be used for the selection of the members?
16. Explain the advantages and disadvantages of using a consulting committee.
17. What are the advantages and limitations of using personal participation as the base for curriculum decisions?
18. Develop a plan for establishing a consulting committee. Explain how you would use the committee for advice on curriculum.
19. Develop a suitable questionnaire for use in making a job analysis of a selected part of your occupation.
20. Plan an on-site analysis of a new phase of your occupation.
    a. Develop the procedure to follow.
    b. Structure the forms for recording information.
    c. Translate the findings into outline form.

REFERENCES

1. Cay, Donald F., *Curriculum: Design for Learning.* Indianapolis: The Bobbs-Merrill Company, Inc., 1966.
2. *Guide for Analyzing Jobs* (Analyst's Workbook). Washington: U.S. Department of Labor, 1966.
3. *Instructional Analysis of the Automotive Mechanic Occupation.* Tallahassee: Division of Vocational, Technical, and Adult Education, State Department of Education, May, 1967.
4. *Instructional Analysis of the Cosmetology Occupation.* Tallahassee: Division of Vocational, Technical, and Adult Education, State Department of Education, September, 1968.
5. Jeffrey, Ruby P., and Fred Tancy, *Job Evaluation Systems Informational Manual.* Phoenix: Arizona State Employment Service, 1967.
6. Lessinger, Leon M., "Educational Stability in an Unstable Technical Society," *Journal of Secondary Education,* May, 1965.
7. Morrison, Edward J., *Development and Evaluation of an Experimental Curriculum for the New Quincy (Mass.) Vocational-Technical Schools*

(U.S. Office of Education Project No. 5–0009, Contract No. OE–5–85–019). Pittsburgh: American Institutes for Research, June 30, 1965, Appendix B.

8. Morsh, J. E., "Job Analysis." Paper presented to the National Society of Programmed Instruction, San Antonio, March 29, 1963.
9. Nelsen, Arden, and Ronald Westfall, *Training and Reference Manual for Job Analysis,* BES No. #–E–3. Washington: U.S. Department of Labor, 1965.
10. *The Richmond Plan.* Report of a study by the Richmond (California) Union High School District and Cogswell Polytechnical College. San Francisco: Cogswell Polytechnical College, 1963.
11. Riendeau, Albert J., *The Role of the Advisory Committee in Occupational Education in the Junior College.* Washington: American Association of Junior Colleges, 1967.

Chapter 5 USING ZONED ANALYSIS CHARTS TO GET THE BIG PICTURE

The use of validated information as the basis for occupational analysis was emphasized in Chapter 4. After securing the best possible information on the current requirements of the payroll classification for the job, the next task is to translate the competencies needed for employment into effective education for the student. This requires further analysis. The delimitation of the identified requirements of the payroll job or occupation to suit the objectives possible in the educational program is the next logical step.

Certain controls are imposed on every educational program. Some of these are: amount of time, available equipment, resources for learning and teaching, background of the instructor, and similar factors. In order to achieve the maximum outcome for the student within the controls imposed, further analysis is desirable to translate opportunities for learning into preparation for entry-level jobs. The task of decision making is simplified by effective use of organizational charts. Two of these are the zoned analysis chart and the content analysis chart. The use of the zoned analysis chart will be explained and illustrated in this chapter.

Purpose and Function of Zoned Analysis Charts

The zoned analysis chart is an instrument for graphically illustrating the results of analysis. The function is to portray the division of the whole into parts and show the resulting relationships. Many subjects may be illustrated by zoned analysis. For example, an occupation can be subdivided to show payroll classifications and the duties of individuals in these payroll classifications. This method of graphic delineation lends itself very well to the portrayal of the organization of programs of instruction. A "bird's-eye view" can be readily obtained by using the zoned analysis of the division of the curriculum into courses and of courses into units.

The zoned analysis approach is a method of programming relationships which can be readily adapted to any kind of organizational structure or system. It may be used to illustrate such diverse structures as school–community relationships, topics to be presented at a meeting, or the outline of the main points of a speech.

The zoned analysis really becomes a master plan of the activities or organization for which it was developed. It logically shows the breakdown of the general concepts into increasing degrees of specificity. The zoned analysis chart is in the form of a series of circles circumscribed around a central core. Each of the circles establishes the boundaries of the zone between the circles. The number of zones will vary with the nature of the project and the degree of refinement of the desired analysis.

HOW TO MAKE A ZONED ANALYSIS

Each of the zones may be identified by a particular title. The center or primary area is the *core* and contains the title of the subject being analyzed. The zone immediately surrounding the core contains areas identified as the *fields.* The fields in turn are in a zone adjacent to an outer zone containing the *branches.* In like manner, each zone may be used to contain additional subdivisions of the larger zone. As each zone may constitute the base for finer subdivisions, so each zone may also be split into two or more parts. Sometimes an informational area is used between two zones or between the core and zone to describe the next breakdown or provide other essential information to aid in the interpretation of the zoned analysis chart.

The construction of the zoned analysis chart is simplified if it is considered in relation to a three-point outline. Assume that a topic for outlining has been selected. The title is then determined. Subsequently, the main section of the outline is then identified and each main section is subdivided into smaller units. These smaller units in turn may be subdivided into still smaller parts and so on until the degree of detail desired is secured. In Figure 5–1 the parts of a three-point outline are shown in broad categories and related to the zoned analysis chart.

In the construction of the zoned analysis chart the title is superimposed upon the core of the smallest circle. The space between the circle of the core and the second circle constitutes the first zone. This zone is divided proportionately between the number of Roman numerals, in this case two. So, two fields are indicated. The first field, I, has three main parts. Each of these becomes a part or branch of the field. In this case the first

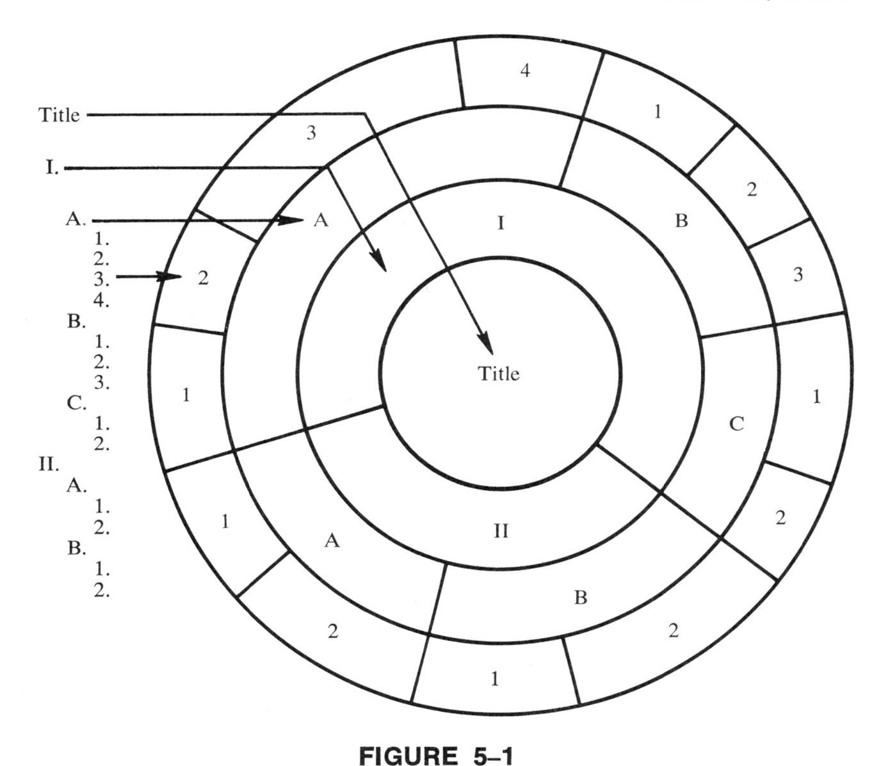

FIGURE 5–1

THREE-POINT OUTLINE SHOWN IN RELATION TO A ZONED ANALYSIS CHART

branch is also divided into three smaller parts or divisions. In like manner the entire outline is transposed into the zoned analysis chart.

A graphic delineation of what zoned analysis is and how it is used is shown in Figure 5–2.

Further explanation of the process of production and reading of the chart is provided in Figure 5–3.

The basic process of construction of the zoned analysis chart has been illustrated in Figures 5–1, 5–2, and 5–3. The plan consists of listing in the core the title of the subject being analyzed. The main divisions or blocks are identified in the first zone with further subdivisions in successive zones as progression continues out from the core. For quick reference each of the zones can be numbered with number one being closest to the core. Each zone may also be given a title, which will restrict the kind of information to be included in that zone.

Partition. The zoned analysis chart should be so constructed that the radial lines separating the first zone into two or more sections extend

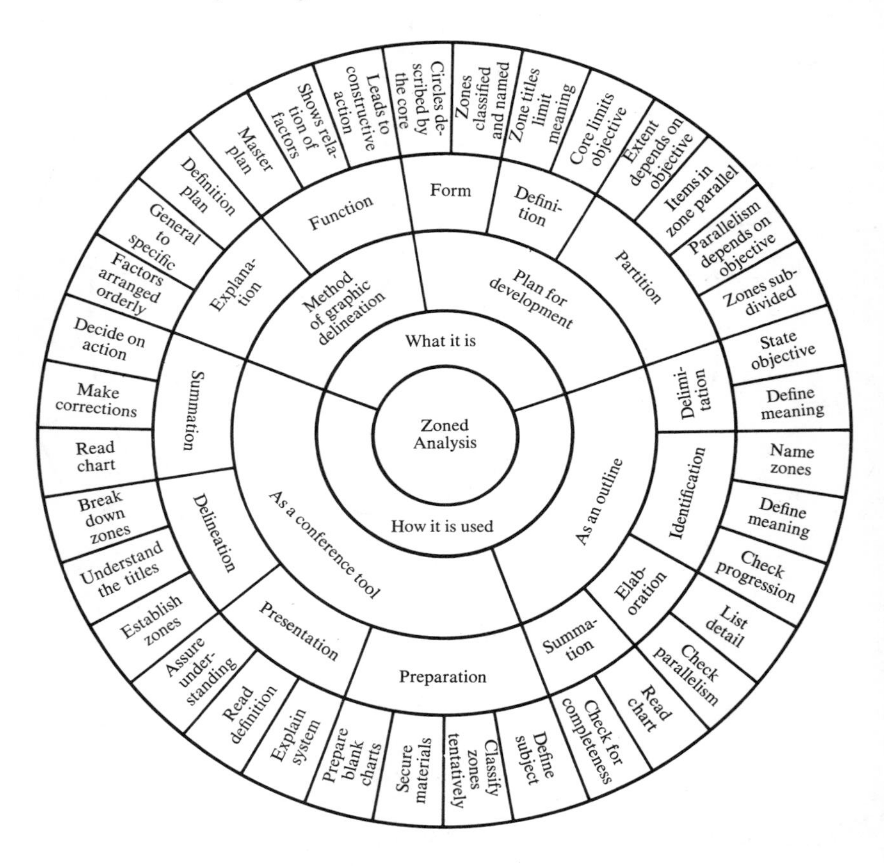

FIGURE 5–2

Explanation of Zoned Analysis by Use of a Zoned Analysis Chart

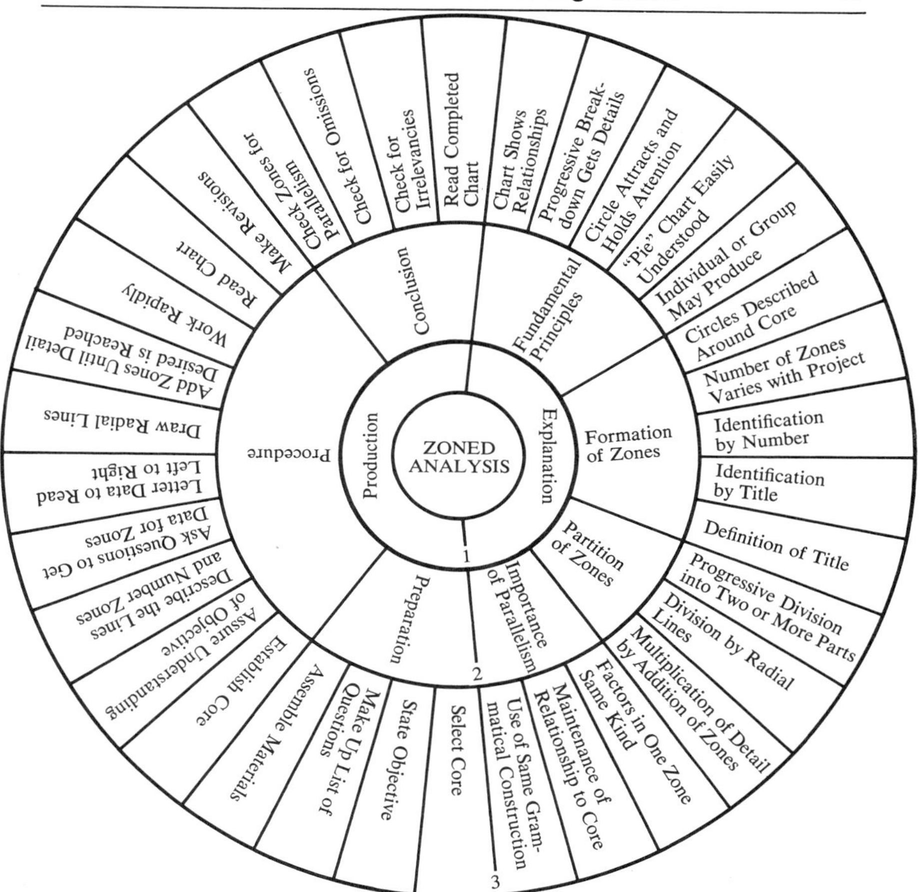

PROCEDURE. Establish the core. This means writing it within a circle and, if necessary, defining the words used. The next step is to read the objective which has been prepared and make sure that it is clearly understood. Describe boundary lines for two or three zones and number them. Then proceed by considering a question which will lead to a breakdown of the first zone. Ask question one, stated above, and decide whether the average vocational educator would want an explanation and tips on production of a *Zoned Analysis*. Enter the items selected so they can be read from left to right on the right side of the chart. It is now a simple matter to draw in the radial lines and proceed to the next zone. It is a good plan to proceed rather rapidly; avoid belaboring an issue. Most of the points will be clarified as the analysis develops.

Read the chart occasionally by adding transitional words. For example, "To explain *Zoned Analysis,* I must state fundamental principles involved, the factors involved in a formation of zones, and the factors affecting partition of zones. I must impress the people with the importance of parallelism."

Discrepancies show up reading the chart from the core outward and again inward to the core.

Source: A Unit of Instruction. Denver: Denver Public Schools, 1962, p. 45.

FIGURE 5–3

ZONED ANALYSIS CHART: EXPLANATION AND PRODUCTION

through all zones from the exterior core to the periphery of the extreme exterior zone. Each item in the first zone breaks down into two or more parts in zone 2. In like manner each part in zone 2 subdivides into two or more parts in zone 3. Should additional zones be used the same principle would apply. The result is a chart with clearly indicated relationships and greater specificity as the exterior extremities are approached.

Parallelism. Each part of the same zone should be of the same nature and equality with every other part of that zone. This is not difficult if the first approach is to develop a three-point outline from the laundry list of categories of assembled facts. Parallelism will occur in the zoned analysis chart if parallelism is present in the outline. Naturally, the objectives and specifications for each zone must be considered in arriving at the decision as to the grouping of content for each of the zones. Failure to maintain parallelism results in a distorted relationship in the chart.

Readability. The user of the chart will find it more readable if each of the captions is so inserted that the chart can be read without turning the page. Figure 5–2 is so constructed, while Figure 5–3 necessitates the turning of the page in order to read some of the captions.

APPLICATIONS TO OCCUPATIONAL CURRICULUMS

The analysis of the occupation provides information essential to the construction of a zoned analysis of the total curriculum. If the unit method of instruction is used, a simple zoned analysis showing the division of the curriculum into courses and the division of the courses into units is highly desirable. Division of the units into lessons may be shown on the chart with even further breakdowns if desired. If the units of instruction are identified by the zoned analysis method, further detailed breakdown can be better shown on the content analysis chart, which will be discussed in the following chapter. The objectives identified are major concerns reflected in the analysis made and the graphic delineations constructed.

A functional fundamental zoned analysis chart for program planning is shown in Figure 5–4. Radial lines need to be drawn in at appropriate locations to designate the area of each zone identified as separate courses and units of instruction.

Effective zoned analysis charts have been developed for each of several vocational education curriculums. Study of the charts shown in Figures 5–5 through 5–8 will help individuals desiring to make such an analysis chart proceed in a logical manner. Careful study of the charts

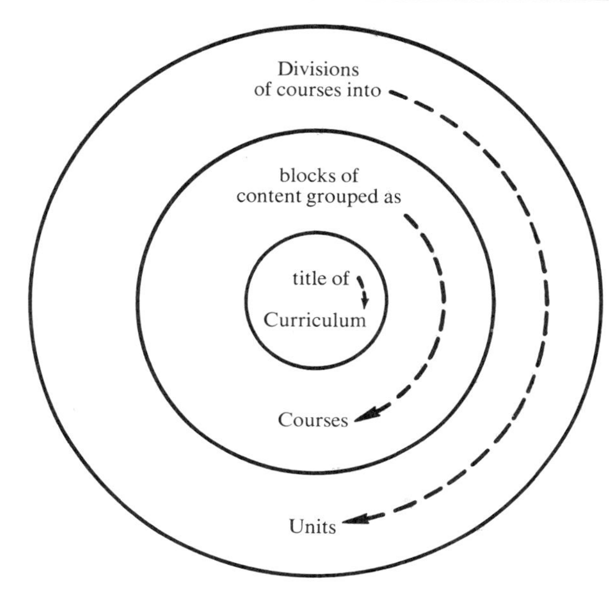

FIGURE 5–4

A FUNDAMENTAL ZONED ANALYSIS FOR PROGRAM PLANNING

will reveal a variety of approaches that may be taken as well as suggest varying solutions to the problems.

Each chart needs to grow out of a validated determination of the manpower needs in that particular occupation or family of payroll jobs.

The reader may wish to take each chart and study the development of the zones. A logical approach is to read each chart beginning with the core adding the first zone within radial lines, then moving into the second zone selecting the content between two appropriate radial lines and proceeding in this manner until the extreme peripheral line has been reached. It will be necessary in many cases to add a few words to make the reading smooth. Always return to the core and commence the next sentence.

The zoned analysis charts, Figures 5–5 through 5–8, are occupational zoned analysis. In Figure 5–5 the work performed by mechanical drafting technicians is shown. Analysis of the work performed in industry by the auto mechanic technician is depicted in the chart, Figure 5–6. The payroll function performed by television service technicians is given in Figure 5–7. Figure 5–8 provides information about the woodworking industry.

(Text continued on p. 84)

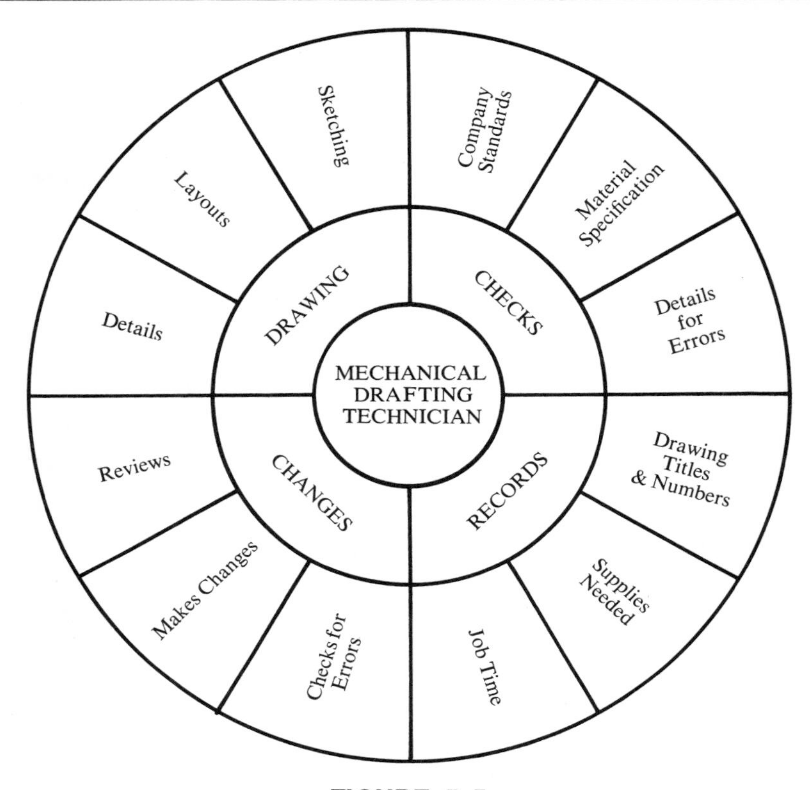

FIGURE 5–5

MECHANICAL DRAFTING TECHNICIAN ZONED ANALYSIS CHART

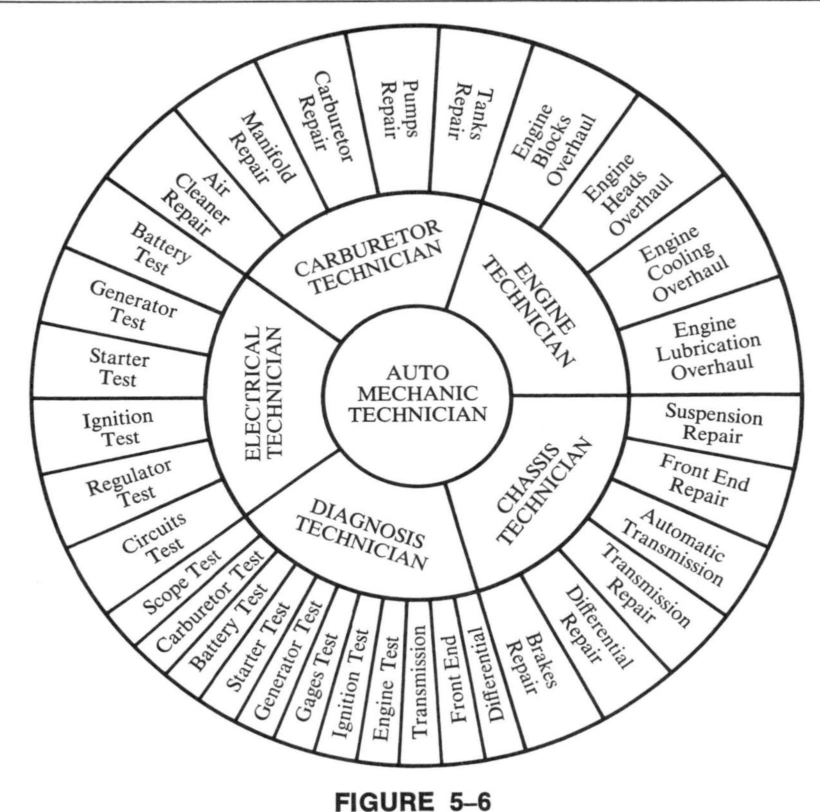

FIGURE 5–6

AUTO MECHANIC TECHNICIAN ZONED ANALYSIS CHART

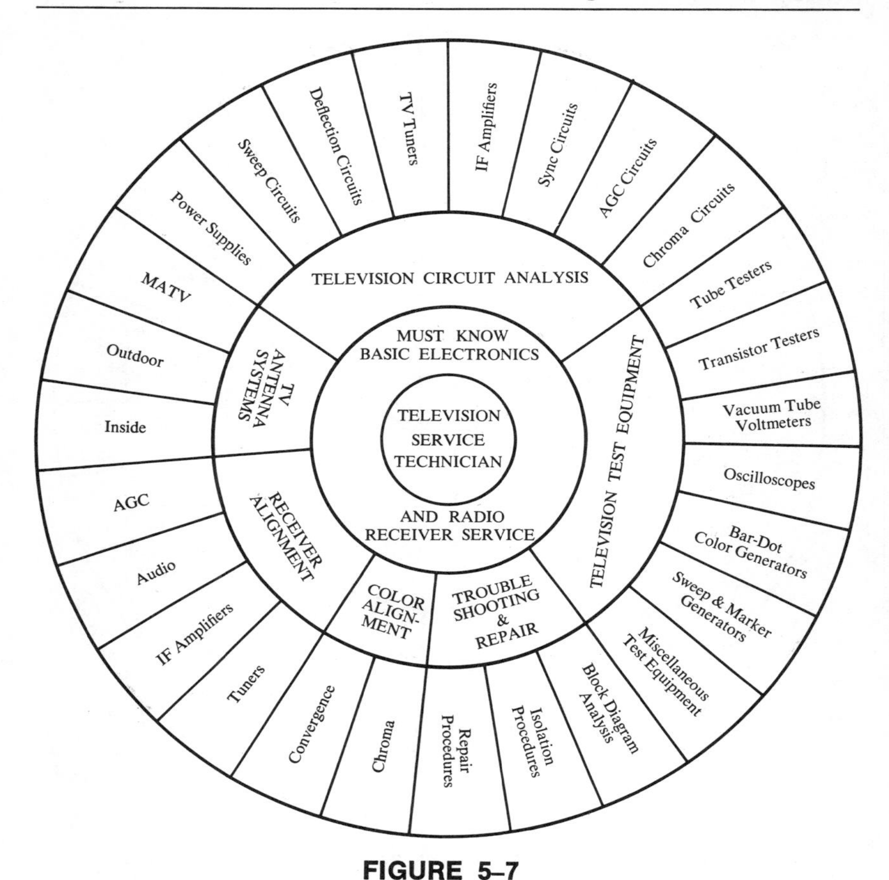

FIGURE 5–7

TELEVISION SERVICE TECHNICIAN ZONED ANALYSIS CHART

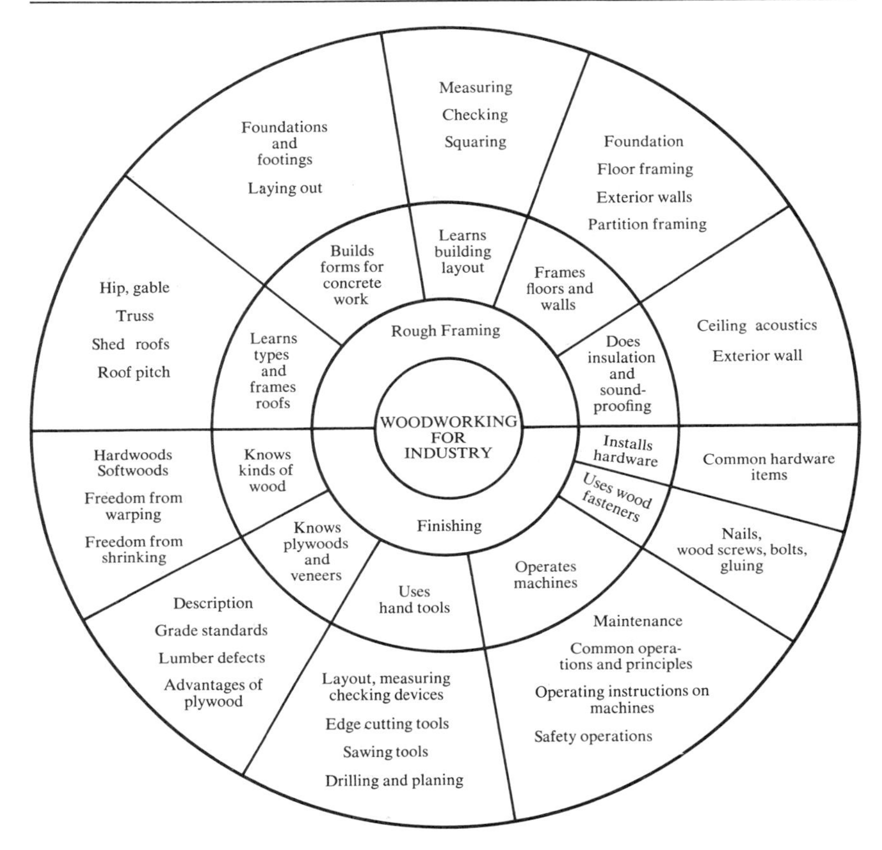

FIGURE 5–8

WOODWORKING INDUSTRY ZONED ANALYSIS CHART

OTHER APPLICATIONS

The zoned analysis chart is adaptable to many uses. Figures 5–9 and 5–10 illustrate some of these uses.

Figure 5–9 portrays general education in three zones with the major emphasis on socio-civil participation and personal growth.

Figure 5–10 illustrates a zoned analysis of the unit of instruction on the four-step teaching method using the main breakdowns of preparation, presentation, application, and examination.

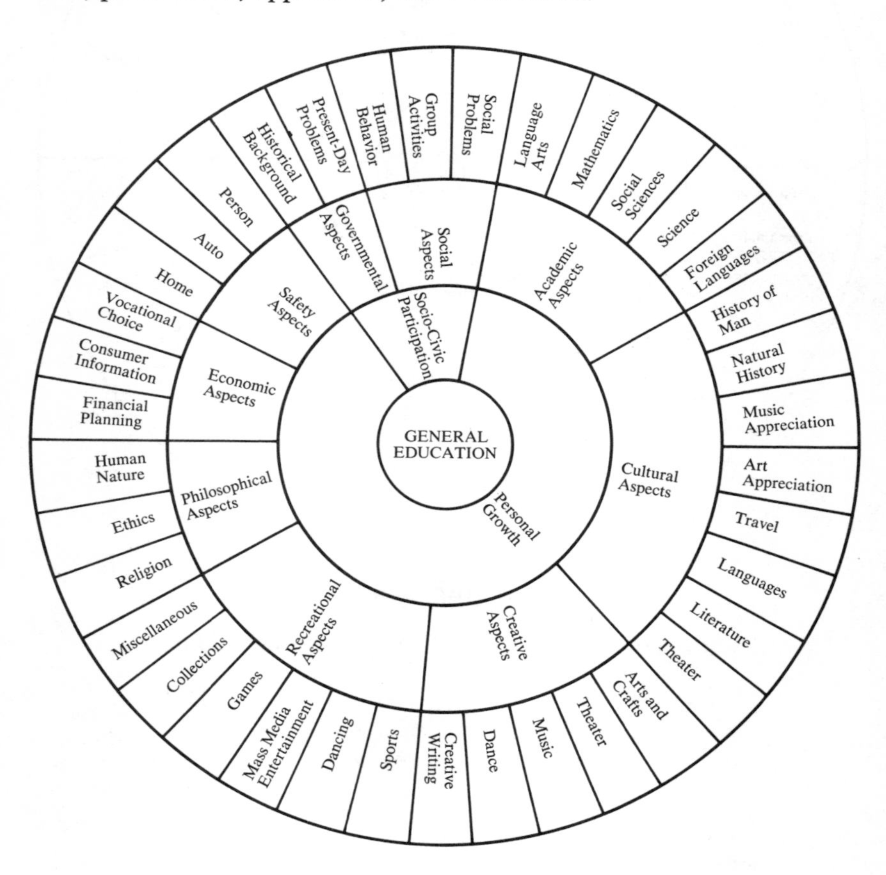

Source: A Unit of Instruction. Denver: Denver Public Schools, 1962.

FIGURE 5–9

GENERAL EDUCATION ZONED ANALYSIS CHART

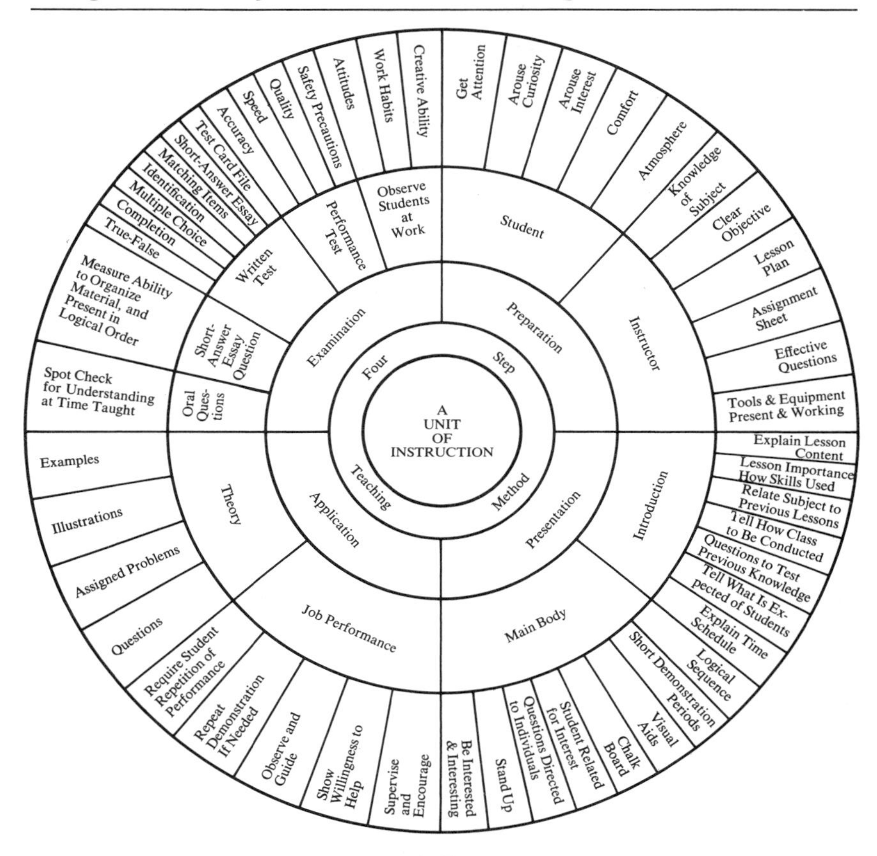

FIGURE 5–10

ZONED ANALYSIS CHART OF A UNIT OF INSTRUCTION

Typical zoned analysis charts developed to show curriculums are provided in Figures 5–11 through 5–14. The curriculum is identified in the core. The next zone may be an information zone, which usually is followed by the division of the curriculum into courses. Reading to the next zone provides the breakdown into units.

Continued breakdown into finer and finer elements is possible by the zoned analysis chart; however, when the units have been determined, the analysis of the unit is more readily accomplished by the use of the content analysis chart, which is discussed in the next chapter.

(Text continued on page 90)

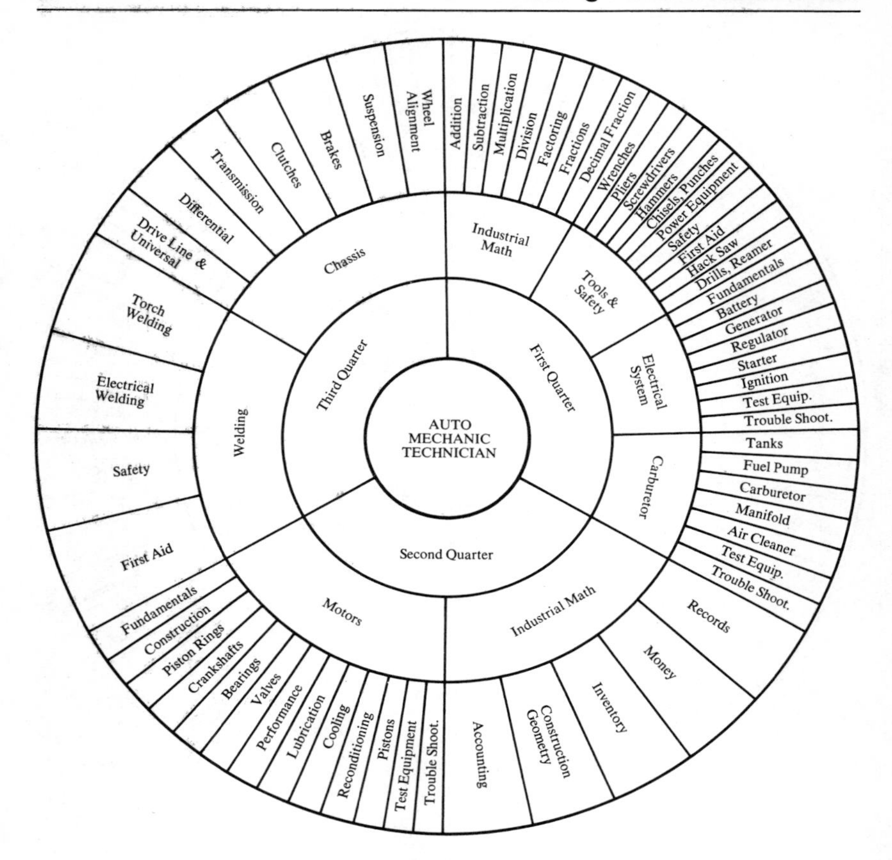

FIGURE 5–11

CURRICULUM ZONED ANALYSIS FOR A TWO-YEAR PROGRAM
(FIRST YEAR)

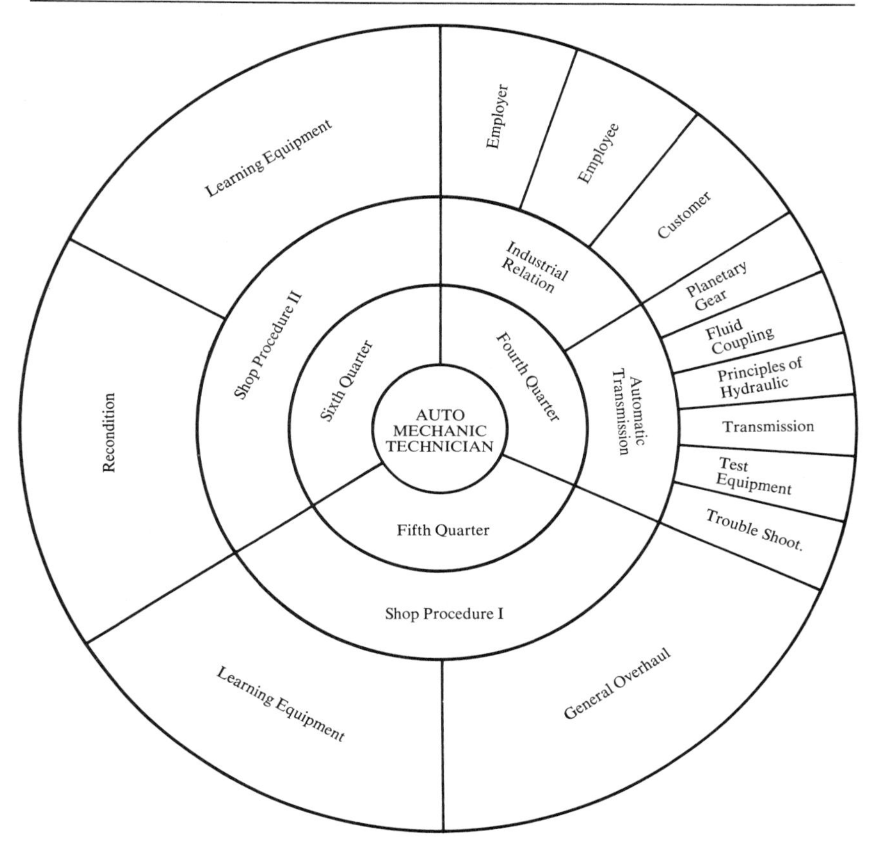

FIGURE 5–12

CURRICULUM ZONED ANALYSIS FOR A TWO-YEAR PROGRAM (SECOND YEAR)

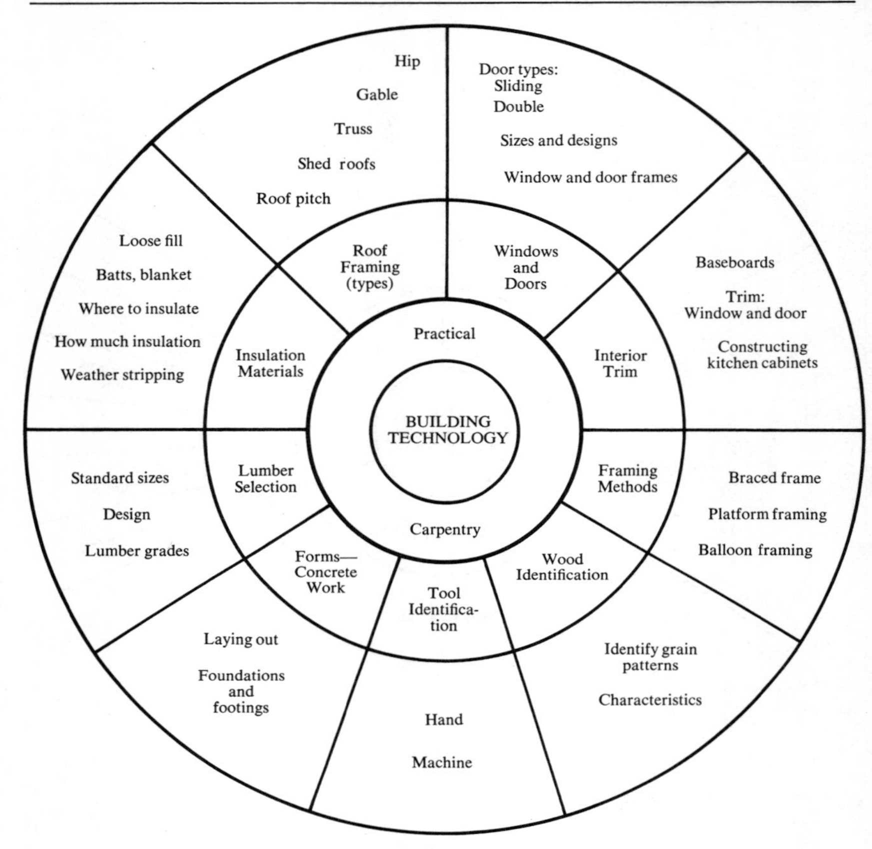

FIGURE 5–13

CURRICULUM ZONED ANALYSIS OF BUILDING TECHNOLOGY

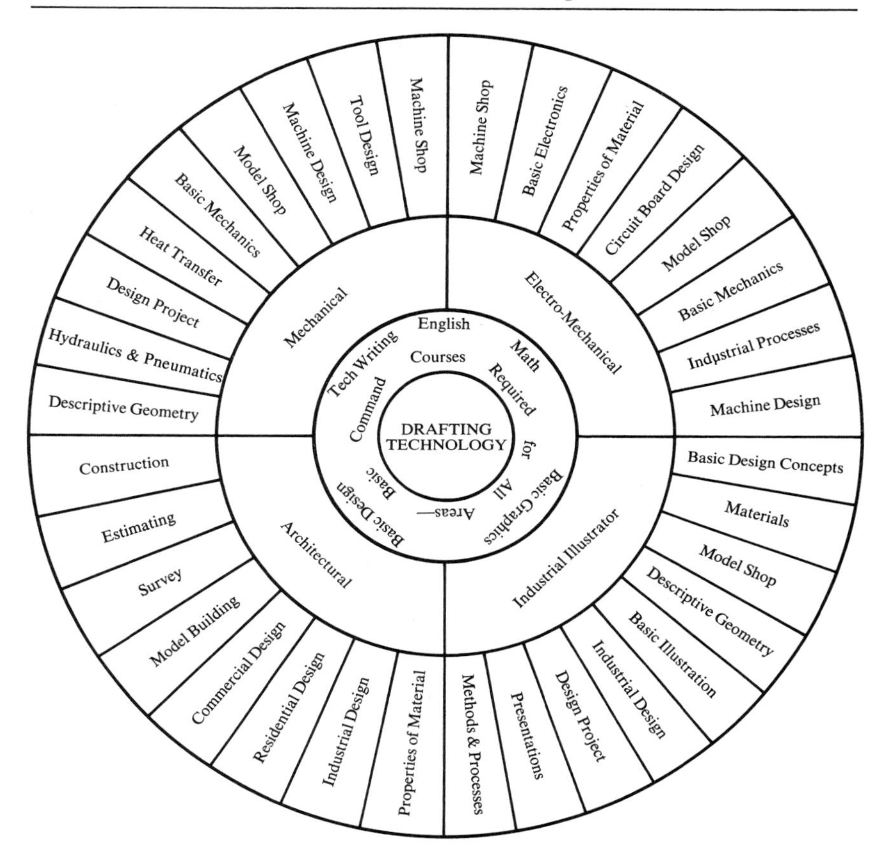

FIGURE 5–14

ZONED ANALYSIS CHART OF THE CURRICULUM OF DRAFTING TECHNOLOGY

Summary

The zoned analysis chart is a simple and useful instrument for gathering into one master plan the most important elements of any organizational structure. It is ideally suited to curricular organizational planning, and the portrayal of payroll job classifications and occupational clusters.

If the zoned analysis chart is to be used with the analysis of occupations, its use will follow the collection of data. The data should be validated by study of original sources if considered necessary. Secondary sources, books, handbooks, pamphlets, and similar materials often provide much of the data for analysis.

Organizing the data into a three-point outline is helpful prior to the construction of the zoned analysis chart. This organization establishes the major blocks and the divisions and subdivisions of the blocks, thus simplifying the task of translating the information into chart form.

The steps of the process of building a zoned analysis are:

1. Determine the title and insert in the circular core.
2. Circumscribe the core with a larger circle to represent the fields.
3. Install radial lines separating the blocks of the field.
4. Circumscribe a large circle around the zone for the fields to provide for the branches.
5. Extend the radial lines through this zone and insert additional radial lines between the parts representing various branches of the zone.
6. Continue until the desired specificity of the zoned analysis has been achieved.
7. Check the chart for: partitions, parallelism, and readability.

QUESTIONS AND ACTIVITIES

1. What is a zoned analysis chart?
2. Explain the advantages of the zoned analysis chart.
3. Translate into a zoned analysis chart the outline previously developed of your occupation.
4. To what part of a three-point outline do the following parts of the zoned analysis chart compare:
   a. Core
   b. Field

   c. Branch
   d. Division

5. What different uses may be made of a zoned analysis chart?
6. Explain the procedure to use in constructing a zoned analysis chart.
7. What is meant by parallelism of a zoned analysis chart?
8. What is the purpose of radial lines on the zoned analysis chart?
9. Demonstrate the technique of reading a zoned analysis chart.
10. Construct a zoned analysis chart of your occupation.
11. Eliminate from the occupational zoned analysis chart content not to be included in the curriculum. Organize the curriculum content into a logical approach and develop as a curriculum content analysis chart.
12. Illustrate how the zoned analysis chart can be used as a conference tool.
13. List several other uses for the zoned analysis chart.
14. Explain the difference between job analysis and curriculum analysis.
15. Delineate between the occupational analysis and the zoned analysis chart.

Chapter 6 CONSTRUCTING THE CONTENT ANALYSIS

In previous chapters the process of obtaining validated information about the paycheck job or occupation was described. Analysis was recognized as the process of dividing and subdividing the whole into parts. This is an essential step in order to derive that content from the job or occupation which must be incorporated into the curriculum. Curriculum building is, in essence, a construction process from the essential parts identified through analysis of the occupation.

To facilitate the illustration of the organizational functions of the analysis process the zoned analysis chart was used. The zoned analysis chart served as a master plan to provide the big picture in a concentrated form. To further expedite an orderly process of translating the detailed plan of analyzed content into instructional materials a content analysis chart is used.

The zoned analysis chart is an excellent vehicle for showing the breakdown of the curriculums into blocks of information which constitute the course in the academic term of the educational institution. The zones on this chart can be used to show the units of instruction into which each of the courses is divided.

Purpose and Function of Content Analysis Charts

As soon as the unit has been identified the content analysis chart can be used effectively to show the vehicles of instructional sheets desirable for a software system of individually paced instruction.

The content analysis chart shows the operations, jobs, and information sheets in coded sequence for effective use in the learning and teaching process.

INCREMENTS OF THE LEARNING STRUCTURE

The increments of the learning structural components must be limited at any one time for most effective learning. If the basic increment is too large for the comprehension of the student, ineffective or very little learning takes place. The size of the increment is influenced by the span of attention of the learner and the length of the instructional period. Mastery of the basic component of the learning system permits combinations with other basic components to form a larger package. These packages with supporting information may then be combined to constitute an integrated instructional system. The effectiveness of the instructional system may be monitored in process and after completion of the learning activities as well as through feedback from individuals having secured payroll jobs. This provides performance evaluation of the learner and the learning process. If the system of learning and teaching is functioning smoothly the results of performance evaluation must coincide or be in phase with the performance goals identified prior to the beginning of the instructional process.

At this time it is highly important to visualize the total structure of the integrated learning system. (See Figure 6–1.) This may be a complex system of computerized instruction or it may be a simple system consisting of:

A. Analysis of the payroll job and/or occupation
B. Determination of behavioral objectives
C. Programming the organizational components
   1. Zoned analysis
   2. Content analysis
      a. Operation sheets
      b. Job sheets
      c. Information sheets
      d. Job plan sheets
      e. Assignment sheets

D. Performance evaluation
   1. Process (The instruction itself)
   2. Product (Feedback from students and employers)

E. Modification of the learning system or process (if necessary)

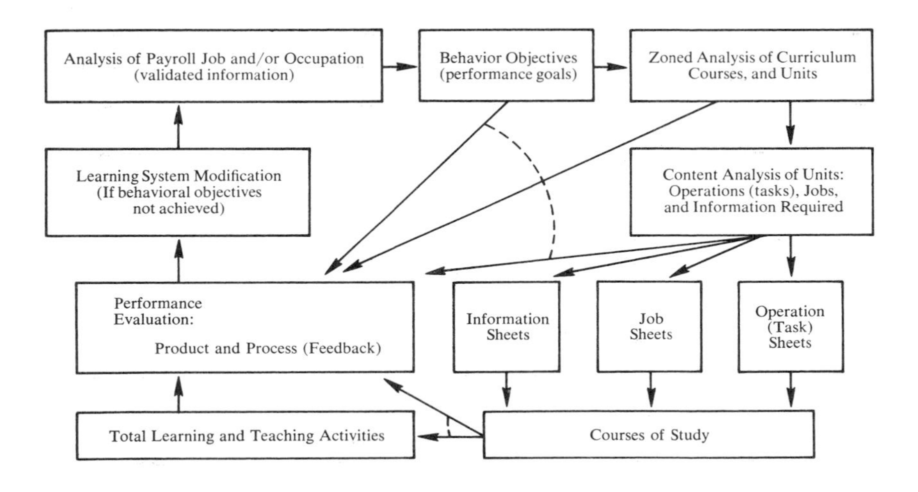

FIGURE 6–1

A SIMPLE INTEGRATED (CLOSED-LOOP) LEARNING SYSTEM FOR A COURSE

OPERATION

An operation is an increment of work of the job, completed project, or training activity. Task and operation are terms descriptive of approximately the same element or size of analytical division of the job. The operation is a convenient instructional increment suited to the period of instructional time and the span of interest of the student.

The operation is a unit of work in a job that involves the making, servicing, or repairing of something. The operation may involve depicting, forming, shaping, assembling, or a similar activity. (1, p. 42)

The operation may be cognitive (mental) as well as psychomotor (manual). Operations are basic skills that must be performed again and again in doing the work of the occupation. (3, p. 50)

STEP

A step is a subdivision of an operation. Each operation must consist of two or more steps. The step represents the smallest part of the analysis sequence commonly employed in the breakdown of the occupation. If the increment of work cannot be divided into smaller units, it is a step. While steps and key points are employed in the operation sheets, at this time it is necessary only to recognize the difference between steps and operations. Key points are important factors related to the steps in the instructional process. This will alert the student to elements of danger, the need for precaution, or the need to emphasize in some way an essential element in the performance of the step. If the key point is not observed the step will not be correctly completed.

JOB

The job is the completed project or training activity. The job is composed of two or more operations (or tasks). These operations are so integrated and organized when successfully performed as to result in a completed product. The completed product may be one that involves manufacturing, repairing, or servicing. The job successfully completed in business, industry, or agriculture results in compensation to the individual.

The job constitutes the secondary rather than the primary increment of instruction in programs of education and/or training. The operation is usually recognized as the basic increment for learning and teaching.

COMPARISON OF JOB, OPERATION, AND STEP

It is necessary to be able to distinguish jobs from operations, and both from steps of operations in the construction of the content analysis chart and instructional materials sheets. The relationship is shown in Figure 6–2.

A simple comparison may be made to the stacking of boxes. Two or more small boxes stacked adjacent within a larger box represent the

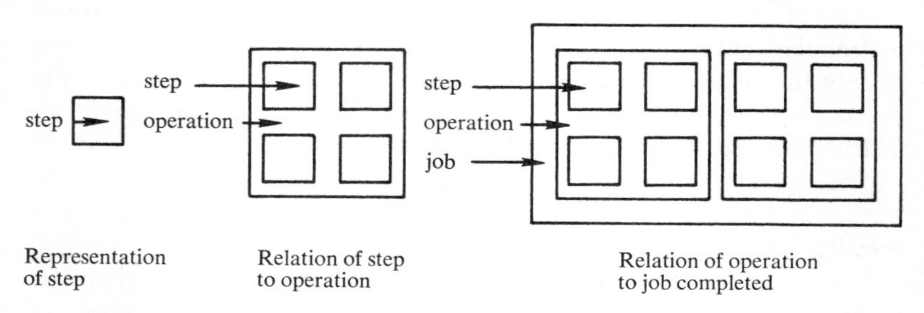

FIGURE 6–2

RELATIONSHIPS OF JOBS, OPERATIONS, AND STEPS

relationship of two or more steps to an operation. Then if two or more of the larger boxes are again placed adjacent to each other within a still larger box, the relationship of operations to the job would be illustrated.

INFORMATION (RELATED OR APPLIED)

Successful completion of a job involves the performance of operations (mental and manual) and the application of information gained concerning the components involved. The information may consist of essential facts for the performance of the operation and/or job. Other types of information may provide background knowledge resulting in more satisfactory or meaningful work experience. Some of the information may be guidance oriented.

The essential information for the performance of the job or operation is often described as *must-know* information; while the other kinds of information are classified as *nice-to-know* information.

Information provided by the shop or laboratory instructor either during the same class or during a separate period is usually called related information. In many post-high school institutions a separate instructor teaches the content of such information in separate applied information classes.

Related information may consist of such topics as kinds of materials, types of tools, methods of fabrication, required mathematics, essential scientific concepts, code requirements, etc. Other topics of related information may identify job opportunities and requirements, opportunities for promotion, factors affecting success on the job, how to apply for positions, etc.

Content Analysis Chart

Several of the components of the software of a simple instructional system are integrated in the content analysis chart for organizational purposes. This chart reflects the applications of the psychology of learning to individual needs in a self-paced system.

Selection has been made of the most desirable and typical jobs for teaching the skills and knowledges involved. The jobs have been broken down into the required operations. The operations have been arranged in logical order. Essential information, to be learned simultaneously with the mastery of the operations in the performance of the job, has been delineated. All of these components have been packaged for learning and teaching with the topics identified on the content analysis chart.

One form for a content analysis chart is shown in Figure 6–3. Space is provided in the title block for identification of the titles of the unit and

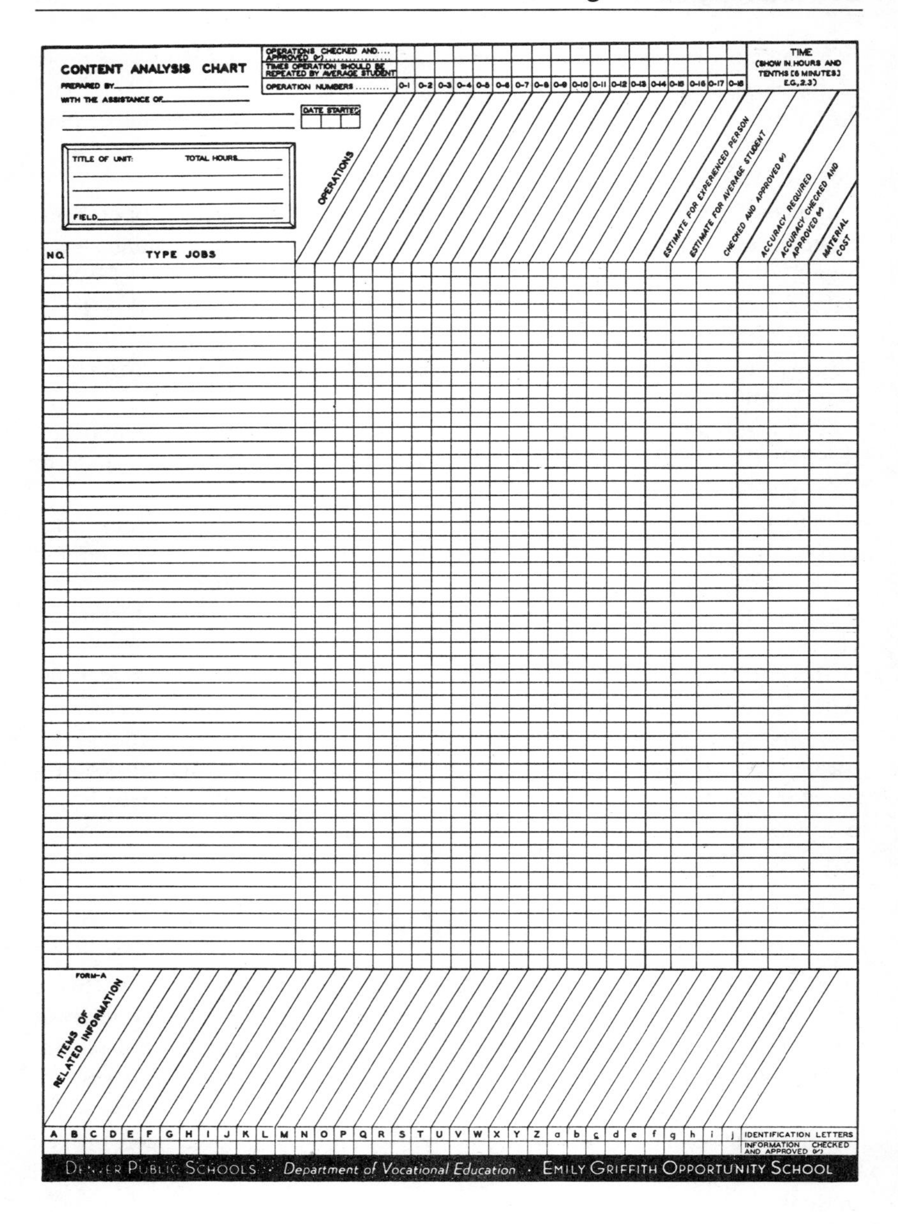

CONTENT ANALYSIS CHART

PREPARED BY________

WITH THE ASSISTANCE OF________

OPERATIONS CHECKED AND APPROVED (✓)

TIMES OPERATION SHOULD BE REPEATED BY AVERAGE STUDENT

OPERATION NUMBERS: O-1, O-2, O-3, O-4, O-5, O-6, O-7, O-8, O-9, O-10, O-11, O-12, O-13, O-14, O-15, O-16, O-17, O-18

TIME (SHOW IN HOURS AND TENTHS [6 MINUTES] E.G., 2.3)

DATE STARTED

TITLE OF UNIT:

TOTAL HOURS________

FIELD________

OPERATIONS

NO.	TYPE JOBS	ESTIMATE FOR EXPERIENCED PERSON	ESTIMATE FOR AVERAGE STUDENT	CHECKED AND APPROVED (✓)	ACCURACY REQUIRED	ACCURACY CHECKED AND APPROVED (✓)	MATERIAL COST

FORM-A

ITEMS OF RELATED INFORMATION

A	B	C	D	E	F	G	H	I	J	K	L	M	N	O	P	Q	R	S	T	U	V	W	X	Y	Z	a	b	c	d	e	f	g	h	i	j	IDENTIFICATION LETTERS
																																				INFORMATION CHECKED AND APPROVED (✓)

DENVER PUBLIC SCHOOLS · Department of Vocational Education · EMILY GRIFFITH OPPORTUNITY SCHOOL

Source: A Unit of Instruction—How to Organize It and How to Teach It. Denver: Denver Public Schools, 1967.

FIGURE 6–3

FORM OF CONTENT ANALYSIS CHART

the field which was determined and inserted on the zoned analysis chart.

Typical jobs of the occupation are selected for teaching. These are listed on the left-hand side of the chart under "Type Jobs." The type jobs are broken down into operations. These operations are placed in logical order for learning at the top of the chart. The process described can be reversed. Based upon information gained from the previous process of analysis the operations are identified. On the basis of the identified operations typical jobs for teaching these operations are selected. The objectives and outcomes should be the same, only the order of procedure has been reversed.

The information sheets (related or applied) are identified, and listed at the bottom of the chart. Such information materials should be developed or compiled as necessary for successful performance and desirable for background knowledge related to the learning activities.

The student as well as the teacher is aided by relating the operation sheets and the information sheets to the jobs by means of a meaningful code. Therefore, a numerical sequence is established for the operation sheets with an alphabetical sequence used to denote the information sheets. It is then a simple matter to indicate in the square on the chart for each typical job the order (by numbers) of required operations and the information sheets (by letters) to be used in the completion of the typical training job.

Space is provided on the chart for indicating other performance specifications helpful in the evaluation of the success of the learning experience.

The content analysis chart becomes a plan of organization for the development of the instructional materials. Individual student learning and individualized teaching are enhanced by preparation in advance of required operation sheets, job sheets, and information sheets.

Several illustrations of the applications of the content analysis chart are provided in Figures 6–4 through 6–7. (*Source: A Unit of Instruction–How to Organize It and How to Teach It.* Denver: Denver Public Schools, 1967, pp. iv, ix, xxiv, and n.a.). These figures show the analysis applications to the fields of automotive repair, welding, commercial food service, and television alignment.

The content analysis chart is very flexible. If individuals are accustomed to thinking in terms of tasks and activities (or other terms) these can be substituted in the chart. The word task could be inserted where operation is printed. Activity could be inserted instead of type job.

If information sheets are not desired the numerical sequence could be retained and the alphabetical sequence dropped. Some individuals use a simple check for the operations required for the performance of the

(Text continued on page 104)

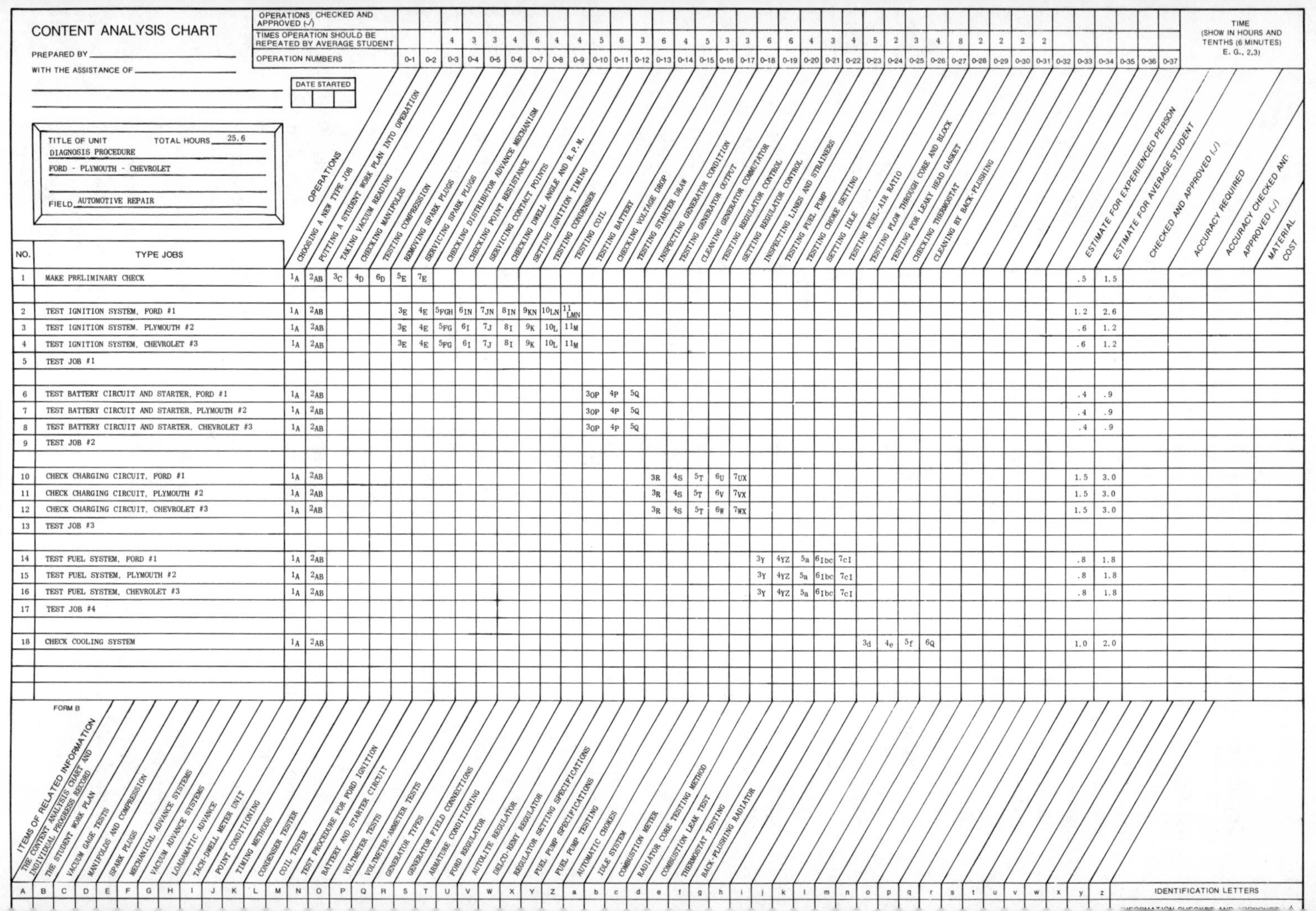

CONTENT ANALYSIS CHART

PREPARED BY ____________

WITH THE ASSISTANCE OF ____________

DATE STARTED

TITLE OF UNIT DIAGNOSIS PROCEDURE
FORD - PLYMOUTH - CHEVROLET
TOTAL HOURS 25.6
FIELD AUTOMOTIVE REPAIR

OPERATIONS CHECKED AND APPROVED (✓)

TIME (SHOW IN HOURS AND TENTHS (6 MINUTES) E. G., 2,3)

OPERATION NUMBERS	0-1	0-2	0-3	0-4	0-5	0-6	0-7	0-8	0-9	0-10	0-11	0-12	0-13	0-14	0-15	0-16	0-17	0-18	0-19	0-20	0-21	0-22	0-23	0-24	0-25	0-26	0-27	0-28	0-29	0-30	0-31	0-32	0-33	0-34	0-35	0-36	0-37
TIMES OPERATION SHOULD BE REPEATED BY AVERAGE STUDENT			4	3	3	4	6	4	4	5	6	3	6	4	5	3	3	6	6	4	3	4	5	2	3	4	8	2	2	2	2						
OPERATIONS	CHOOSING A NEW TYPE JOB	PUTTING A STUDENT WORK PLAN INTO OPERATION	TAKING VACUUM READING	CHECKING MANIFOLDS	TESTING COMPRESSION	REMOVING SPARK PLUGS	SERVICING SPARK PLUGS	CHECKING DISTRIBUTOR ADVANCE MECHANISM	CHECKING POINT RESISTANCE	SERVICING CONTACT POINTS	CHECKING DWELL ANGLE AND R.P.M.	SETTING IGNITION TIMING	TESTING CONDENSER	TESTING COIL	TESTING BATTERY	CHECKING VOLTAGE DROP	TESTING STARTER DRAW	INSPECTING GENERATOR CONDITION	TESTING GENERATOR OUTPUT	CLEANING GENERATOR COMMUTATOR	TESTING REGULATOR CONTROL	SETTING REGULATOR CONTROL	INSPECTING LINES AND STRAINERS	TESTING FUEL PUMP	TESTING CHOKE SETTING	SETTING IDLE	TESTING FUEL-AIR RATIO	TESTING FLOW THROUGH CORE AND BLOCK	CHECKING FOR LEAKY HEAD GASKET	CHECKING THERMOSTAT	CLEANING BY BACK-FLUSHING						

NO.	TYPE JOBS	0-1	0-2	0-3	0-4	0-5	0-6	0-7	0-8	0-9	0-10	0-11	0-12	0-13	0-14	0-15	0-16	0-17	0-18	0-19	0-20	0-21	0-22	0-23	0-24	0-25	0-26	0-27	0-28	0-29	0-30	0-31	ESTIMATE FOR EXPERIENCED PERSON	ESTIMATE FOR AVERAGE STUDENT	CHECKED AND APPROVED (✓)	ACCURACY REQUIRED	ACCURACY CHECKED AND APPROVED (✓)	MATERIAL COST
1	MAKE PRELIMINARY CHECK	1A	2AB	3C	4D	6D	5E	7E																									.5	1.5				
2	TEST IGNITION SYSTEM, FORD #1	1A	2AB				3E	4E	5FGH	6IN	7JN	8IN	9KN	10LN	11LMN																		1.2	2.6				
3	TEST IGNITION SYSTEM, PLYMOUTH #2	1A	2AB				3E	4E	5FG	6I	7J	8I	9K	10L	11M																		.6	1.2				
4	TEST IGNITION SYSTEM, CHEVROLET #3	1A	2AB				3E	4E	5FG	6I	7J	8I	9K	10L	11M																		.6	1.2				
5	TEST JOB #1																																					
6	TEST BATTERY CIRCUIT AND STARTER, FORD #1	1A	2AB													3OP	4P	5Q															.4	.9				
7	TEST BATTERY CIRCUIT AND STARTER, PLYMOUTH #2	1A	2AB													3OP	4P	5Q															.4	.9				
8	TEST BATTERY CIRCUIT AND STARTER, CHEVROLET #3	1A	2AB													3OP	4P	5Q															.4	.9				
9	TEST JOB #2																																					
10	CHECK CHARGING CIRCUIT, FORD #1	1A	2AB																3R	4S	5T	6U	7UX										1.5	3.0				
11	CHECK CHARGING CIRCUIT, PLYMOUTH #2	1A	2AB																3R	4S	5T	6V	7VX										1.5	3.0				
12	CHECK CHARGING CIRCUIT, CHEVROLET #3	1A	2AB																3R	4S	5T	6W	7WX										1.5	3.0				
13	TEST JOB #3																																					
14	TEST FUEL SYSTEM, FORD #1	1A	2AB																					3Y	4YZ	5a	6Ibc	7cI					.8	1.8				
15	TEST FUEL SYSTEM, PLYMOUTH #2	1A	2AB																					3Y	4YZ	5a	6Ibc	7cI					.8	1.8				
16	TEST FUEL SYSTEM, CHEVROLET #3	1A	2AB																					3Y	4YZ	5a	6Ibc	7cI					.8	1.8				
17	TEST JOB #4																																					
18	CHECK COOLING SYSTEM	1A	2AB																										3d	4e	5f	6Q	1.0	2.0				

FORM B

ITEMS OF RELATED INFORMATION

IDENTIFICATION LETTERS	ITEM
A	THE CONTENT ANALYSIS CHART AND INDIVIDUAL PROGRESS RECORD
B	THE STUDENT WORK PLAN
C	VACUUM GAGE TESTS
D	MANIFOLDS AND COMPRESSION
E	SPARK PLUGS
F	MECHANICAL ADVANCE SYSTEMS
G	VACUUM ADVANCE SYSTEMS
H	LOADMATIC ADVANCE
I	TACH-DWELL METER UNIT
J	POINT CONDITIONING
K	TIMING METHODS
L	CONDENSER TESTER
M	COIL TESTER
N	TEST PROCEDURE FOR FORD IGNITION
O	BATTERY AND STARTER CIRCUIT
P	VOLTMETER TESTS
Q	VOLTMETER-AMMETER TESTS
R	GENERATOR TYPES
S	GENERATOR FIELD CONNECTIONS
T	ARMATURE CONDITIONING
U	FORD REGULATOR
V	AUTOLITE REGULATOR
W	DELCO-REMY REGULATOR
X	REGULATOR SETTING SPECIFICATIONS
Y	FUEL PUMP SPECIFICATIONS
Z	FUEL PUMP TESTING
a	AUTOMATIC CHOKES
b	IDLE SYSTEM
c	COMBUSTION METER
d	RADIATOR CORE TESTING METHOD
e	COMBUSTION LEAK TEST
f	THERMOSTAT TESTING
g	BACK-FLUSHING RADIATOR
h	
i	
j	
k	
l	
m	
n	
o	
p	
q	
r	
s	
t	
u	
v	
w	
x	
y	
z	

FIGURE 6–4

CONTENT ANALYSIS CHART

PREPARED BY ____________

WITH THE ASSISTANCE OF ____________

DATE STARTED

TITLE OF UNIT: ARC WELDING FLAT POSITION PART I

TOTAL HOURS 138.6

FIELD ELECTRIC WELDING

	O-1	O-2	O-3	O-4	O-5	O-6	O-7	O-8	O-9	O-10	O-11	O-12	O-13	O-14	O-15	O-16	O-17	O-18						
OPERATIONS CHECKED AND APPROVED (✓)																								
TIMES OPERATION SHOULD BE REPEATED BY AVERAGE STUDENT				8	10	12	6	6	8	6	9	8	7	9	9	7	11							
OPERATION NUMBERS	O-1	O-2	O-3	O-4	O-5	O-6	O-7	O-8	O-9	O-10	O-11	O-12	O-13	O-14	O-15	O-16	O-17	O-18	TIME (SHOW IN HOURS AND TENTHS (6 MINUTES) E.G., 2.3)					
TYPE JOBS / OPERATIONS	CHOOSING A NEW TYPE JOB	FILLING IN A STUDENT WORK PLAN	PUTTING STUDENT WORK PLAN INTO OPERATION	ADJUSTING WELDING MACHINE	MAKING A STRINGER BEAD	TYING IN A BEAD	MAKING A WEAVE BEAD	PADDING	FITTING A LAP JOINT FOR WELDING	TACKING A LAP JOINT FOR WELDING	WELDING A LAP JOINT	FITTING AN OPEN CORNER JOINT	TACKING AN OPEN CORNER JOINT	WELDING AN OPEN CORNER JOINT	FITTING A "T" JOINT	TACKING A "T" JOINT	WELDING A "T" JOINT		ESTIMATE FOR EXPERIENCED PERSON	ESTIMATE FOR AVERAGE STUDENT	CHECKED AND APPROVED (✓)	ACCURACY REQUIRED	ACCURACY CHECKED AND APPROVED (✓)	MATERIAL COST
MAKE 16 PASSES ON 1/4" x 5 1/4" x 6 1/4" PLATE	1ACO	2CO	3CO	4BD	5E	6F													1.5	48.0*		± 1/32"		
MAKE 16 PASSES ON 1/4" x 5 1/4" x 6 1/4" PLATE	1ACO	2CO	3CO	4BD	5E	6F													1.5	5.0		± 1/32"		
MAKE 16 PASSES ON 1/4" x 5 1/4" x 6 1/4" PLATE	1ACO	2CO	3CO	4BD	5E	6F													1.5	3.0		± 1/32"		
MAKE 16 PASSES ON 1/4" x 5 1/4" x 6 1/4" PLATE	1CO	2CC	3CO	4BD	5E	6F													1.5	1.5		± 1/32"		
MAKE 10 PASSES ON 3/8" x 5 1/4" x 6 1/4" PLATE	1CO	2CO	3CO	4BD		6F	5H												1.0	10.0*		± 1/16"		
MAKE 10 PASSES ON 3/8" x 5 1/4" x 6 1/4" PLATE	1CO	2CO	3CO	4BD		6F	5H												1.0	1.0		± 1/16"		
BUILD UP 4 PADS 1/4" x 4" x 5" ON 2 3/8" x 5" x 6" PLATES	1CO	2CO	3CO	4BD		6F		5G											3.2	7.1		+ 1/4"		
BUILD UP PAD ON 3/8" x 4" x 4" PLATE	1CO	2CO	3CO	4BD		6F		5G											2.0	3.5		+ 1/4"		
TEST #1																								
WELD 1/4" AND 3/8" x 3" x 6" LAPPED DECK JOINT	1O	2O	3O						4K	5JN	6LI								1.0	20.0*		+ 1/16"		
WELD 3/8" x 1" x 6" STIFFNER TO 3/8" PLATE	1O	2O	3O						4K	5JN	6LI								1.0	2.5		+ 1/16"		
WELD TWO 3/8" x 3" DIAMETER PADS TO 3/8" PLATE	1O	2O	3O						4K	5JN	6LI								0.8	2.0		+ 1/32"		
WELD TWO 1/4" x 2" x 3" SHIM PADS TO 3/8" PLATE	1O	2O	3O						4K	5JN	6LI								1.0	1.5*		+ 1/16"		
WELD TWO 1/4" x 2" DIAMETER PADS TO 1/4" PLATE	1O	2O	3O						4K	5JN	6LI								0.5	0.5		+ 1/32"		
TEST #2																								
WELD CORNER OF TANK MADE FROM 1/4" PLATE	1O	2O	3O									4KD	5JN	6LI					0.5	1.5		+ 1/16"		
MAKE CHANNEL BEAM SECTION	1O	2O	3O									4KD	5JN	6LI					0.8	2.0		+ 1/16"		
WELD CORNER OF TANK, 3/8" PLATE 8" LENGTH	1O	2O	3O									4KD	5JN	6LI					1.0	1.5		+ 1/16"		
MAKE SPACER BRACKET	1O	2O	3O									4KD	5JN	6LI					2.0	2.0		+ 1/16"		
TEST #3																								
WELD STIFFNER TO FLOOR PLATE 1/4" X 10"	1O	2O	3O												4KD	5JI	6LN		1.0	10.0*		+ 1/32"		
WELD CAP PLATE TO UPRIGHT BEAM	1O	2O	3O												4KD	5JI	6LN		1.0	3.0		+ 1/32"		
WELD CAP PLATE TO ROUND COLUMN	1O	2O	3O												4KD	5JI	6LN		1.0	3.0		+ 1/32"		
WELD 1/4" FOOT PLATE ON 3" PIPE	1O	2O	3O												4KD	5JI	6LN		1.0	2.5		+ 1/32"		
WELD TWO ROUND STOP BOSSES TO PLATE	1O	2O	3O												4KD	5JI	6LN		1.0	2.0		+ 1/32"		
WELD 6 STIFFNER BARS TO PLATE	1O	2O	3O												4KD	5JI	6LN		2.0	3.0		+ 1/32"		
WELD DOUBLE CONNECTING PLATES TO BEAM	1O	2O	3O												4KD	5JI	6LN		1.0	1.5		+ 1/32"		
WELD 30° PIPE FOR BRACE TO FLOOR PLATE	1O	2O	3O												4KD	5JI	6LN		1.0	1.0		+ 1/32"		
TEST #4																								

FORM-A

RELATED INFORMATION

IDENTIFICATION LETTERS	RELATED INFORMATION	INFORMATION CHECKED AND APPROVED (✓)
[illegible]	[illegible] RULES	
B	MACHINE ADJUSTMENTS	
C	THE CONTENT ANALYSIS CHART	
D	ELECTRODE TYPES AND A.W.S. CLASSIFICATIONS	
E	STRINGER BEADS	
F	BEAD TIE-IN	
G	METHODS OF PADDING	
H	WEAVE BEADS	
I	EXPANSION AND CONTRACTION	
J	TACK WELDS	
K	JOINT PREPARATION	
L	FILLET WELDS	
M	GROOVE WELDS	
N	WELD INSPECTION	
O	THE STUDENT WORK PLAN	
P		
Q		
R		
S		
T		
U		
V		
W		
X		
Y		
Z		
a		
b		
c		
d		
e		
f		
g		
h		
i		
j		

FIGURE 6–5

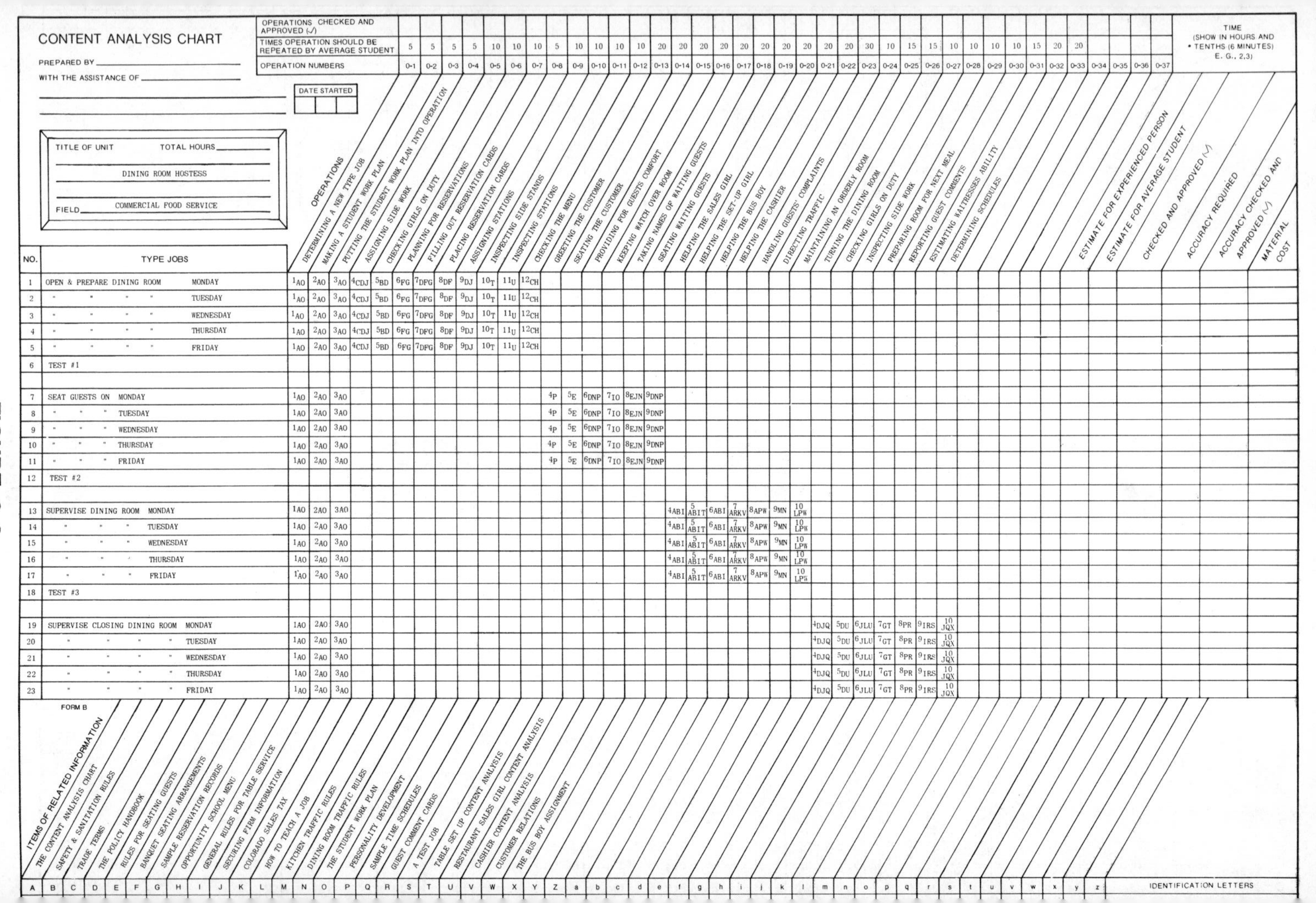

CONTENT ANALYSIS CHART

PREPARED BY ______________

WITH THE ASSISTANCE OF ______________

DATE STARTED

TITLE OF UNIT DINING ROOM HOSTESS   TOTAL HOURS ______

FIELD COMMERCIAL FOOD SERVICE

TIME (SHOW IN HOURS AND • TENTHS (6 MINUTES) E. G., 2,3)

NO.	TYPE JOBS	DETERMINING A NEW TYPE JOB	MAKING A STUDENT WORK PLAN	PUTTING THE STUDENT WORK PLAN INTO OPERATION	ASSIGNING SIDE WORK	CHECKING GIRLS ON DUTY	PLANNING FOR RESERVATIONS	FILLING OUT RESERVATION CARDS	PLACING RESERVATION CARDS	ASSIGNING STATIONS	INSPECTING SIDE STANDS	INSPECTING STATIONS	CHECKING THE MENU	GREETING THE CUSTOMER	SEATING THE CUSTOMER	PROVIDING FOR GUESTS COMFORT	KEEPING WATCH OVER ROOM	TAKING NAMES OF WAITING GUESTS	SEATING WAITING GUESTS	HELPING THE SALES GIRL	HELPING THE SET-UP GIRL	HELPING THE BUS BOY	HELPING THE CASHIER	HANDLING GUESTS' COMPLAINTS	DIRECTING TRAFFIC	MAINTAINING AN ORDERLY ROOM	TURNING THE DINING ROOM	CHECKING GIRLS ON DUTY	INSPECTING SIDE WORK	PREPARING ROOM FOR NEXT MEAL	REPORTING GUEST COMMENTS	ESTIMATING WAITRESSES ABILITY	DETERMINING SCHEDULES						ESTIMATE FOR EXPERIENCED PERSON	ESTIMATE FOR AVERAGE STUDENT	CHECKED AND APPROVED (✓)	ACCURACY REQUIRED	ACCURACY CHECKED AND APPROVED (✓)	MATERIAL COST
	OPERATIONS CHECKED AND APPROVED (✓)																																											
	TIMES OPERATION SHOULD BE REPEATED BY AVERAGE STUDENT	5	5	5	5	10	10	10	5	10	10	10	10	20	20	20	20	20	20	20	20	20	20	30	10	15	15	10	10	10	10	15	20	20										
	OPERATION NUMBERS	0-1	0-2	0-3	0-4	0-5	0-6	0-7	0-8	0-9	0-10	0-11	0-12	0-13	0-14	0-15	0-16	0-17	0-18	0-19	0-20	0-21	0-22	0-23	0-24	0-25	0-26	0-27	0-28	0-29	0-30	0-31	0-32	0-33	0-34	0-35	0-36	0-37						
1	OPEN & PREPARE DINING ROOM MONDAY	1AO	2AO	3AO	4CDJ	5BD	6FG	7DFG	8DF	9DJ	10T	11U	12CH																													
2	" " " " TUESDAY	1AO	2AO	3AO	4CDJ	5BD	6FG	7DFG	8DF	9DJ	10T	11U	12CH																													
3	" " " " WEDNESDAY	1AO	2AO	3AO	4CDJ	5BD	6FG	7DFG	8DF	9DJ	10T	11U	12CH																													
4	" " " " THURSDAY	1AO	2AO	3AO	4CDJ	5BD	6FG	7DFG	8DF	9DJ	10T	11U	12CH																													
5	" " " " FRIDAY	1AO	2AO	3AO	4CDJ	5BD	6FG	7DFG	8DF	9DJ	10T	11U	12CH																													
6	TEST #1																																											
7	SEAT GUESTS ON MONDAY	1AO	2AO	3AO										4P	5E	6DNP	7IO	8EJN	9DNP																									
8	" " " TUESDAY	1AO	2AO	3AO										4P	5E	6DNP	7IO	8EJN	9DNP																									
9	" " " WEDNESDAY	1AO	2AO	3AO										4P	5E	6DNP	7IO	8EJN	9DNP																									
10	" " " THURSDAY	1AO	2AO	3AO										4P	5E	6DNP	7IO	8EJN	9DNP																									
11	" " " FRIDAY	1AO	2AO	3AO										4P	5E	6DNP	7IO	8EJN	9DNP																									
12	TEST #2																																											
13	SUPERVISE DINING ROOM MONDAY	1AO	2AO	3AO																4ABI	5ABIT	6ABI	7ARKV	8APW	9MN	10LPW																		
14	" " " TUESDAY	1AO	2AO	3AO																4ABI	5ABIT	6ABI	7ARKV	8APW	9MN	10LPW																		
15	" " " WEDNESDAY	1AO	2AO	3AO																4ABI	5ABIT	6ABI	7ARKV	8APW	9MN	10LPW																		
16	" " " THURSDAY	1AO	2AO	3AO																4ABI	5ABIT	6ABI	7ARKV	8APW	9MN	10LPW																		
17	" " " FRIDAY	1AO	2AO	3AO																4ABI	5ABIT	6ABI	7ARKV	8APW	9MN	10LPW																		
18	TEST #3																																											
19	SUPERVISE CLOSING DINING ROOM MONDAY	1AO	2AO	3AO																							4DJQ	5DU	6JLU	7GT	8PR	9IRS	10JQX											
20	" " " " TUESDAY	1AO	2AO	3AO																							4DJQ	5DU	6JLU	7GT	8PR	9IRS	10JQX											
21	" " " " WEDNESDAY	1AO	2AO	3AO																							4DJQ	5DU	6JLU	7GT	8PR	9IRS	10JQX											
22	" " " " THURSDAY	1AO	2AO	3AO																							4DJQ	5DU	6JLU	7GT	8PR	9IRS	10JQX											
23	" " " " FRIDAY	1AO	2AO	3AO																							4DJQ	5DU	6JLU	7GT	8PR	9IRS	10JQX											

FORM B

IDENTIFICATION LETTERS	ITEMS OF RELATED INFORMATION
A	THE CONTENT ANALYSIS CHART
B	SAFETY & SANITATION RULES
C	TRADE TERMS
D	THE POLICY HANDBOOK
E	RULES FOR SEATING GUESTS
F	BANQUET SEATING ARRANGEMENTS
G	SAMPLE RESERVATION RECORDS
H	OPPORTUNITY SCHOOL MENU
I	GENERAL RULES FOR TABLE SERVICE
J	SECURING FIRM INFORMATION
K	COLORADO SALES TAX
L	HOW TO TEACH A JOB
M	KITCHEN TRAFFIC RULES
N	DINING ROOM TRAFFIC RULES
O	THE STUDENT WORK PLAN
P	PERSONALITY DEVELOPMENT
Q	SAMPLE TIME SCHEDULES
R	GUEST COMMENT CARDS
S	A TEST JOB
T	TABLE SET UP CONTENT ANALYSIS
U	RESTAURANT SALES GIRL CONTENT ANALYSIS
V	CASHIER CONTENT ANALYSIS
W	CUSTOMER RELATIONS
X	THE BUS BOY ASSIGNMENT
Y	
Z	
a	
b	
c	
d	
e	
f	
g	
h	
i	
j	
k	
l	
m	
n	
o	
p	
q	
r	
s	
t	
u	
v	
w	
x	
y	
z	

FIGURE 6–6

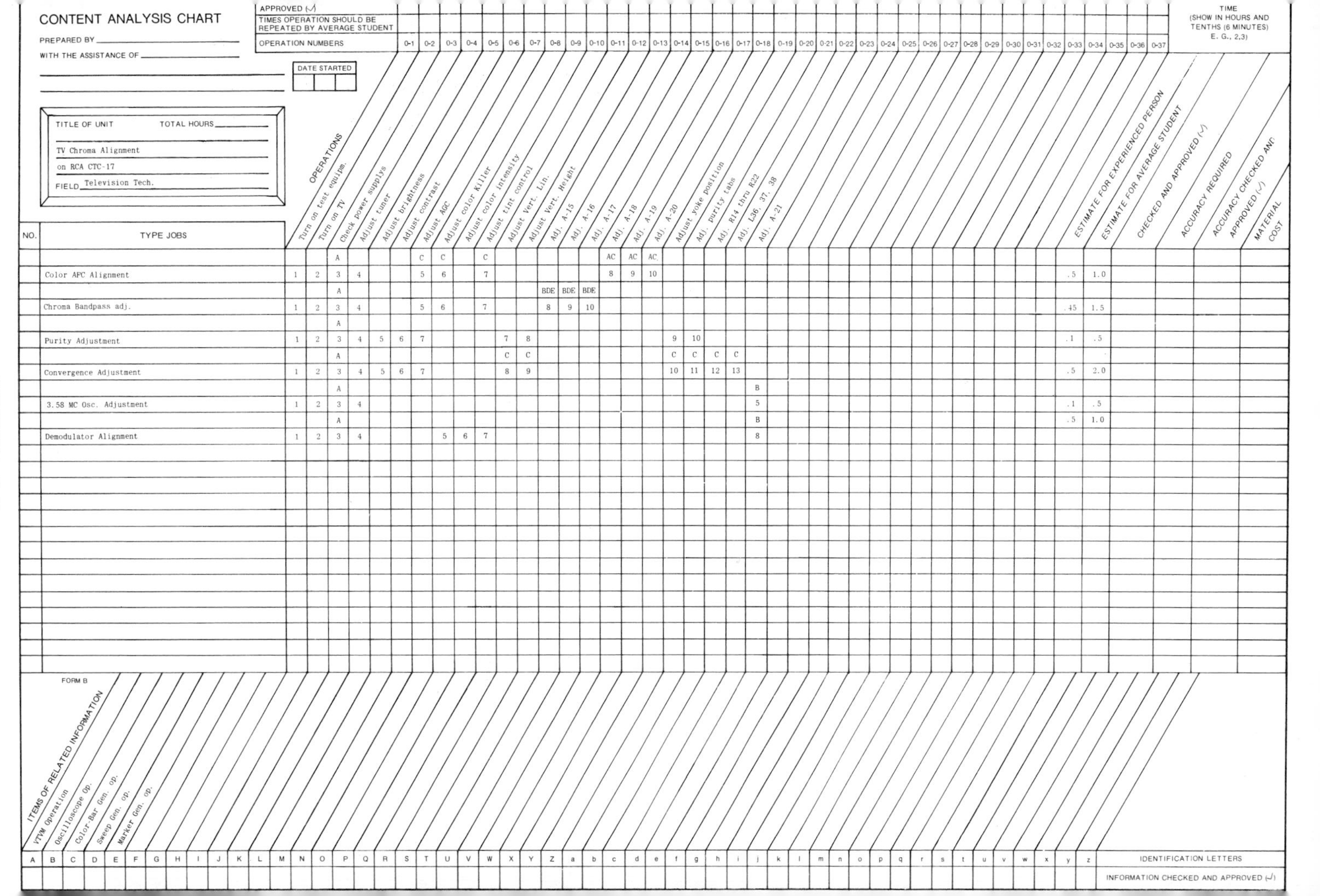

CONTENT ANALYSIS CHART

PREPARED BY ____________

WITH THE ASSISTANCE OF ____________

APPROVED (✓)

TIMES OPERATION SHOULD BE REPEATED BY AVERAGE STUDENT

OPERATION NUMBERS: 0-1 0-2 0-3 0-4 0-5 0-6 0-7 0-8 0-9 0-10 0-11 0-12 0-13 0-14 0-15 0-16 0-17 0-18 0-19 0-20 0-21 0-22 0-23 0-24 0-25 0-26 0-27 0-28 0-29 0-30 0-31 0-32 0-33 0-34 0-35 0-36 0-37

TIME (SHOW IN HOURS AND TENTHS (6 MINUTES) E. G., 2,3)

DATE STARTED

TITLE OF UNIT TV Chroma Alignment on RCA CTC-17 — TOTAL HOURS ____

FIELD Television Tech.

OPERATIONS

NO.	TYPE JOBS	Turn on test equipm.	Turn on TV	Check power supplys	Adjust tuner	Adjust brightness	Adjust contrast	Adjust AGC	Adjust color Killer	Adjust color intensity	Adjust tint control	Adjust Vert. Lin.	Adjust Vert. Height	Adj. A-15	Adj. A-16	Adj. A-17	Adj. A-18	Adj. A-19	Adj. A-20	Adjust yoke position	Adj. purity tabs	Adj. R14 thru R22	Adj. L36, 37, 38	Adj. A-21	ESTIMATE FOR EXPERIENCED PERSON	ESTIMATE FOR AVERAGE STUDENT	CHECKED AND APPROVED (✓)	ACCURACY REQUIRED	ACCURACY CHECKED AND APPROVED (✓)	MATERIAL COST
				A				C	C		C						AC	AC	AC											
	Color AFC Alignment	1	2	3	4			5	6		7						8	9	10						.5	1.0				
				A										BDE	BDE	BDE														
	Chroma Bandpass adj.	1	2	3	4			5	6		7			8	9	10									.45	1.5				
				A																										
	Purity Adjustment	1	2	3	4	5	6	7				7	8							9	10				.1	.5				
				A								C	C							C	C	C	C							
	Convergence Adjustment	1	2	3	4	5	6	7				8	9							10	11	12	13		.5	2.0				
				A																				B						
	3.58 MC Osc. Adjustment	1	2	3	4																			5	.1	.5				
				A																				B	.5	1.0				
	Demodulator Alignment	1	2	3	4				5	6	7													8						

FORM B

ITEMS OF RELATED INFORMATION

A	B	C	D	E
VTVM Operation	Oscilloscope Op.	Color-Bar Gen. op.	Sweep Gen. op.	Marker Gen. op.

IDENTIFICATION LETTERS: A B C D E F G H I J K L M N O P Q R S T U V W X Y Z a b c d e f g h i j k l m n o p q r s t u v w x y z

INFORMATION CHECKED AND APPROVED (✓)

FIGURE 6–7

job rather than the numerical sequence. While this may be somewhat simpler it does not identify the order of performance of operations.

The content analysis chart can be used as a program for learning by the student and as a program of instruction by the teacher.

Typical Jobs for Training

Just as the operation is the basic increment for learning and teaching so the job is the vehicle for translating in-school activities to employment-type experiences. Relevant learning experiences constitute the basis for motivation in school. Interest and motivation stimulate the learner to greater effort. The experiences in school must provide a controlled challenge—one that is similar to the work expected on the payroll job but still within the comprehension and ability of the student.

The jobs selected for the instructional program must be typical of the jobs which the student will later need to do when he secures employment. In addition, the jobs must be selected for the maximum training value and the suitability for the school environment. Four factors are significant to consider in the selection of jobs for learning and teaching. (2, pp. 165–66) These are:

1. Jobs should be selected on the basis of their applicability.
2. Jobs should be selected for the specific skills and knowledge involved. Adequate repetition must be provided.
3. Each job should be evaluated as to its relative difficulty. The student should move from the known to the unknown. Jobs should be selected to provide learning of simple operations prior to exposure to more difficult ones. Requirements of speed, accuracy, finish, muscular coordination, knowledge, and judgment should be considered in the selection of training jobs to suit students' level of achievement.
4. Each job should produce a meaningful and useful result. The student should be given the feeling of accomplishment resulting from completion of worthwhile activities. The student should be given responsibility for the completion of the job within acceptable standards of excellence comparable to those of a payroll job as soon and as frequently as possible.

Summary

The content analysis chart provides the structure for the planning and organizing of the principal components of the instructional materials of a simple individually paced system for learning and teaching.

The zoned analysis chart is helpful in making the determination of the curriculum, courses, and units of instruction. The content analysis chart continues the analysis process beginning with the unit. The unit includes operations, jobs, and information sheets. Essential operations are determined. Topics of information are identified and developed and typical jobs requiring these operations are selected to supplement content readily available in texts and reference books.

The operation is the basic increment of the learning process. Each operation will consist of two or more steps. Steps are the smallest element of the learning process used in the method. The job will consist of two or more operations.

Terms of similar meaning		Increment
step—increment	=	smallest division
operation—task	=	unit of learning
job—activity	=	completed project

Information may be related or applied. It may also be classified as must-know, nice-to-know, and guidance-type information.

The content analysis chart provides the source of information for the development of operation sheets, job sheets, and information sheets. These instructional sheets permit the student to progress at his own rate. They also permit the teacher to work with several different levels of students in the same group without wasting time and without confusion.

The content analysis chart is a valuable aid in the development of a system of instruction suited to the needs of all students.

QUESTIONS AND ACTIVITIES

1. Explain the function of a content analysis chart.
2. How does the content analysis chart grow out of the zoned analysis chart of the curriculum?
3. What are the characteristics of an individually paced learning system?
4. Contrast the features of a "hardware" system of individualized instruction with a "software" system.
5. What is the basic ingredient of the "software" system discussed in this book?

6. Explain the relationships between:
   a. Step and operation
   b. Operation and job
7. What is a key point? When should it be indicated?
8. What are the advantages of a closed-loop software individualized learning system? Disadvantages?
9. Establish essential criteria for the selection of suitable jobs for the units of instruction.
10. Explain the coding system of the content analysis chart.
11. Why should estimated time be determined for performance of the job by the student?
12. Construct a content analysis chart showing one or more instructional units identified from the zoned analysis chart.
13. Why is the unit a good organizational structure for teaching and learning?
14. What is meant by:
    a. Related information?
    b. Applied information?
    c. Must-know information?
    d. Nice-to-know information?
    e. Guidance-type information?
15. Are the zoned and content analysis approaches suitable to all instructional services of vocational and technical education? Explain.

REFERENCES

1. Fryklund, Verne C., *Occupational Analysis: Techniques and Procedures.* New York: The Bruce Publishing Company, 1970.
2. *The Preparation of Occupational Instructors.* Washington, D.C.: U.S. Office of Education, 1966.
3. *A Unit of Instruction—How to Organize It and How to Teach It.* Denver: Denver Public Schools, 1967.

Part II CONSTRUCTING THE CURRICULUM

Chapter 7 PLANNING THE RELATED CURRICULUM AND THE COURSES OF STUDY

A curriculum is an integrated sequence of courses of appropriate type and length arranged in a logical order and designed to achieve the defined goals of education. Valid curriculum is the result of analysis. Analysis of the occupation translated into a program of action to prepare students for functional responsibility in the world of work suggests the process as well as the product of an integrated vocational and technical program of learning. Curriculum development is fundamental and essential to an effective program.

Purpose and Scope of Curriculum

The curriculum becomes the organizational structure for program planning. It is the road map which indicates where the students need to go and how to get there. The purpose of the curriculum is to provide the student with a relevant learning experience. Satisfactory completion of the curriculum should result in the competencies essential for employment at entry- or higher-level positions in the field of study.

The curriculum must reflect a balanced educational diet designed with the purposes for which the program is offered. Education has many purposes. These purposes must be established in perspective, maintained in focus, and assessed against the performance potential of the student who has completed the educational program.

Program planning and curriculum building constitute the design stage of the educational enterprise. Much of the success or failure of the student can be traced directly to the design for learning—the curriculum.

The curriculum, according to C. Thomas Olivo:

> affects the nature of the physical plant, the shops, the laboratories, and the related classrooms;
> circumscribes the size, type, and specifications for equipment, instruments, materials, etc.;
> identifies the qualities and other requirements of students/trainees seeking occupational training and education;
> determines the qualifications of the professional teaching, supervisory, and administrative staff; and
> influences every facet of administration, supervision, instruction, guidance, follow-up, and the like. (4, pp. 44 and 117)

The curriculum must correlate the occupational analysis and the imperative educational needs with the instructional program. In 1968 the Advisory Council on Vocational Education identified the following as imperative educational needs significant for curriculum development:

> No longer can the emphasis be on matching the best man with an existing job; it must be placed on providing a suitable job for each man or equipping the man to fill a suitable job.
>
> Less emphasis must be placed on manpower as an economic resource and more on employment as a source of income and status for workers and their families.
>
> A reorientation of values is needed to satisfy a new set of closely interwoven functions.
>
> The opportunity must be provided to improve the individual's employment status and earnings and to help him adapt to a changing economic environment and an expanding economy.
>
> Career consciousness must be integrated throughout the schools in order to enlarge the number of options and alternatives for individual pupils—both in terms of occupations and higher education.
>
> The study of the world of work is a valid part of education for all children—it documents for youth the necessity of education both academic and vocational. (5)

Fundamental concepts for curriculum are the following selected principles of educational philosophy:

> Our first task, a task to which everything else is subordinate, is that of making American citizens; and therefore before we begin

> to specialize too closely in vocational education we should provide a firm basic educational foundation.
>
> Let us apply our finest educational insight and courage to the need for vocational education, insisting that the educated workman is the most valuable of national assets, and the nation which possesses this asset will be a successful competitor.
>
> No vocational school (or program) can turn out a finished journeyman, but it can develop the materials out of which a finished journeyman can be made.
>
> Vocational education programs should be under the control of public schools with representatives of labor on the school board.
>
> All vocational programs should be open to all; sex, creed, color or nationality should not debar anyone.
>
> In order to establish vocational education on a firm and lasting basis, interests of employee and employer must be equally considered.
>
> Stimulus, financial aid, and backing of the government are necessary in order to give vocational education character, direction, and uniformity.
>
> Both general and technical education are important as a means of prevention of the waste of human resources.
>
> Vocational education must deal with adolescents, taking them as they come and fitting them for the practical tests of social and industrial efficiency.
>
> Society cannot continue to expend vast sums of money for high schools and universities and neglect the ninety percent of the students who go into vocational life improperly prepared, without repudiating the reasons usually given for having schools of any sort as a public charge. (1, p. 8)

The concepts referred to indicate another definition of curriculum. In a broad sense this reflects a definition of curriculum expressed by Olivo:

> Whenever the term "curriculum" is used, it refers to the sum-total of all the planned teaching and learning experiences in a controlled situation. Conditions are provided so that the learner may master effectively the skills and technologies for employment (in a cluster of jobs within an occupational constellation) together with common learnings. This combination of experiences and knowledge within the field of specialization and general education represents a vocational curriculum which insures ultimate success as a worker and citizen. (3, p. 3)

It is obvious that the word "curriculum" has different meanings and connotations. The scope of the curriculum must reflect the determined objectives of the total training program. This usually consists in programs of vocational and technical education of three main categories:

1. The specialty shop and/or laboratory subjects
2. The related or applied subjects
3. The general education courses

Lynn A. Emerson described the technician curriculum in the following way:

> The curriculum is a scheduled composite of the courses needed to provide a well-organized training program in a specific area. It thus grows out of the needs of the person in the technician occupation. It includes the technical content together with the basic science and mathematics needed, and may contain units of general education character. The courses are of appropriate length and level, arranged in orderly sequence to facilitate ease of learning.
> Curriculums may be developed entirely from occupational analyses, supplemented by appropriate mathematics and science. They are often constructed by piecing together portions of similar curriculums in other institutions. Perhaps the best technique is a combination of occupational analysis and ideas obtained from curriculums in other institutions. (2, p. 3)

MATERIALS INCLUDED IN DESIGNING CURRICULUM

A practical approach to the delineation of content for the total curriculum of vocational and technical education is to identify the skills and knowledges (as well as habits and attitudes) that the student must possess to perform on the payroll jobs. It is then necessary to subtract that which the student already knows, and that which he can more effectively acquire on the job if it can be postponed until after employment. This is illustrated in Figure 7–1.

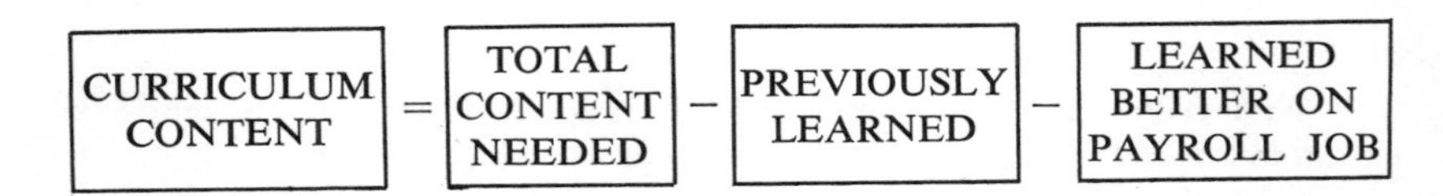

FIGURE 7–1

CONTENT TO INCLUDE IN CURRICULUM

The curriculum materials include all the written and audiovisual aids used by the teacher and student to achieve the objectives identified for the curriculum.

While the primary purpose of curriculum materials is to aid the student in mastering effectively and efficiently the skills, knowledges, attitudes, and habits for which education and training are provided, they do have other uses. Some of their other important uses are:

1. To aid teachers and industrial trainers in performing their function.
2. To assist supervisors and administrators in evaluating, assessing, and coordinating the success of the program.
3. To help teacher educators in the preparation of new teachers and the updating of teachers currently employed.

CURRICULUM DESIGN PRINCIPLES

A person who designs curriculum for vocational and technical education is in a role similar to the designer of a product for business and industry. The user of the product must be considered. Likewise, attention must be given to providing the quality essential for the purpose intended. A systems design of curriculum construction is illustrated in Figures 7–2 and 7–3. Briefly, some of the more important principles of curriculum design include:

1. Consideration of established specifications of need.
2. Identification of the objectives for the "target" audience.
3. Determination of a meaningful curriculum title and one which creates desire to participate in the program.
4. Delimitation of curriculum content to the level of the program, and the ability of the students.
5. Selection of curriculum content that lends itself to school instruction.
6. Consideration of the available equipment.
7. Recognition of established entrance requirements.

CURRICULUM DEVELOPMENT PROCESS

The process of curriculum development involves several steps since the curriculum serves as a master plan. Curriculum in this frame of reference is used to mean an integrated group of courses rather than the total educational experience of the student. These include:

1. Selecting content from analysis of the job or occupation.
2. Setting objectives for the total curriculum.
3. Setting entrance requirements for admittance to curriculum.
4. Recognizing curriculum controls.
5. Grouping content into blocks for courses.
6. Allocating content among selected courses.
7. Considering clustering content.
8. Determining length of courses.
9. Identifying instructional order.
10. Planning offerings of courses by school terms.

11. Deciding types of instruction for each of the courses.
12. Planning time schedule of courses offered.
13. Evaluating effectiveness of instructional program.
14. Considering modifications and revisions suggested to improve the curriculum.

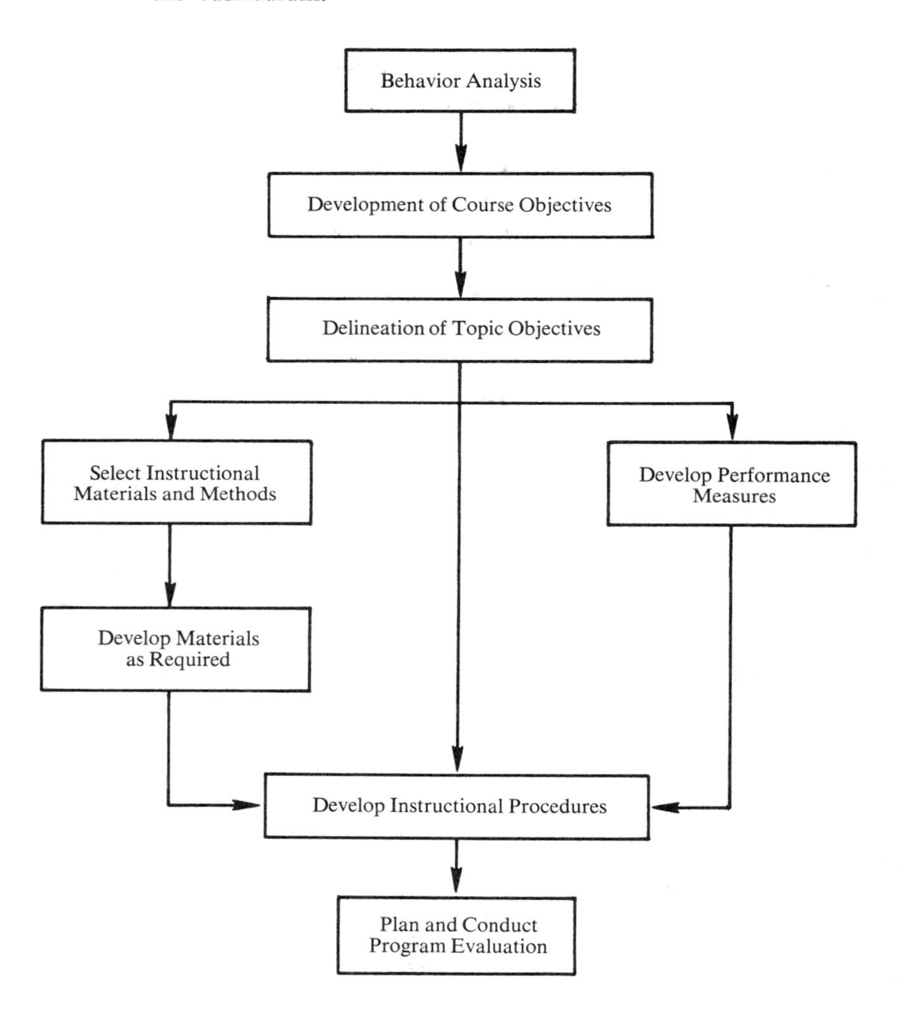

Source: Adapted from Gagné, Robert M., "Development and Evaluation of an Experimental Curriculum for the New Quincy (Mass.) Vocational-Technical School." Pittsburgh: American Institute for Research, 1965.

FIGURE 7–2

SUMMARY STEPS OF THE CURRICULUM DEVELOPMENT PROCESS

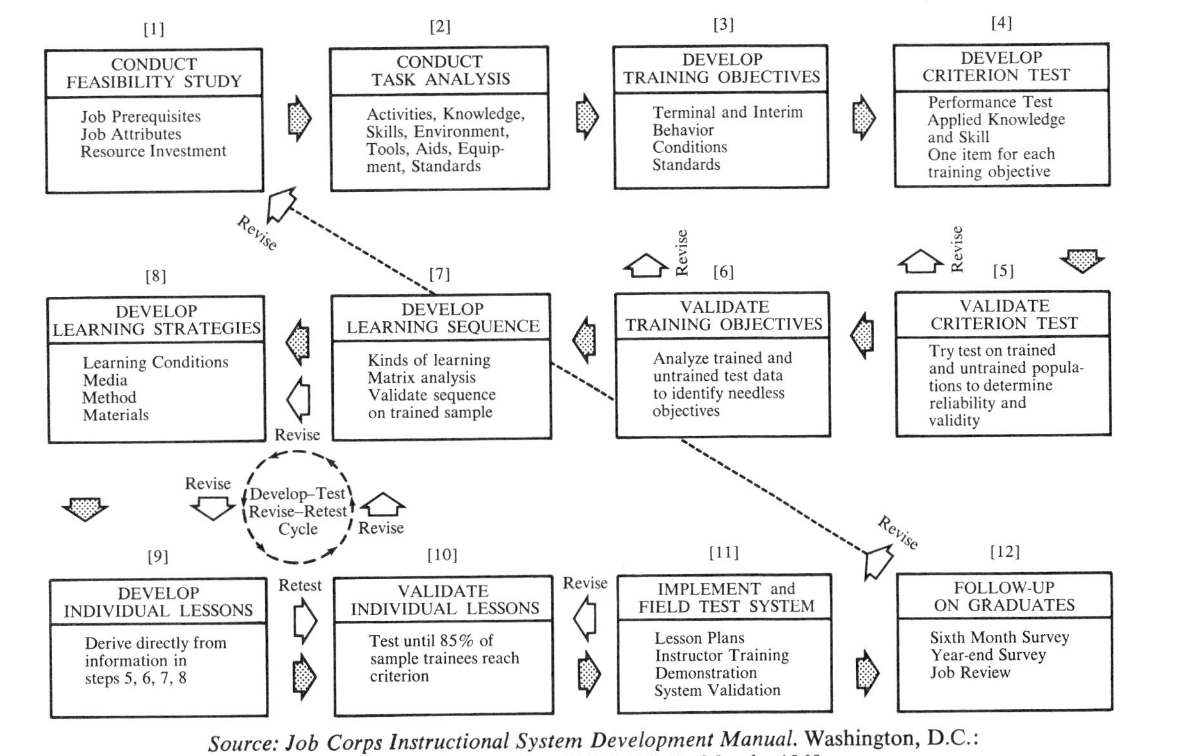

Source: Job Corps Instructional System Development Manual. Washington, D.C.: U.S. Office of Economic Opportunity (JCH 440.4), March, 1968.

FIGURE 7–3

FLOWCHART OF INSTRUCTIONAL SYSTEM DEVELOPMENT PROCESS

CURRICULUM CONTROLS

The effectiveness of the instructional program and success in achieving the objectives are affected by the curriculum controls which exist. Some of these controls are the result of the organizational structure, while others result from the attitudes of the teachers, students, and administrators in the school. Curriculum designs are affected by such elements of curriculum control as the following:

1. Total permissible length of the program.
2. Length of periods for shop or laboratory.
3. Number of available teaching stations in the shop or laboratory; amount of available equipment.
4. Number of courses required of all students. This reduces the time available for the specialty subjects.
5. Number of total hours of general education required of all students.
6. Number of elective courses desired in the curriculum. This limits the number of courses devoted to the specialty.
7. Entrance requirements of the educational institution as a whole.
8. Other requirements for graduation. These may restrict the time or emphasis that can be devoted to the specialty.

METHODS OF BUILDING CURRICULUM

Many methods are employed in building curriculum. Some of these are more valid than others in providing the educational experience needed by the students seeking employment upon the completion of the educational program. These methods are often used in combination depending upon the conditions under which the curriculum builder has to work and the time and resources available to him to complete the process. Briefly the common methods as described by Emerson are:

1. Occupational or job analysis. This is essential for a new occupation or one for which a validated analysis is not available. It is also necessary in order to reflect the recent changes within the job or occupation.
2. "Scissors and paste pot" method. This consists of selecting and using parts of the curriculums of other existing institutions. Those parts which seem desirable are adopted while the remainder is rejected. While this can be done in much less time than the occupational analysis the validity of the product is uncertain.
3. "Lifting" the curriculum from some other institution. This is a highly questionable procedure. The institution may have different objectives, standards, entrance requirements, and place-

ment opportunities. It is even possible that the curriculum was not a good one for that institution. Usually this is the least desirable method to use in building a curriculum for a new institution.

4. Combination of above. (2, p. 40)

CURRICULUM OUTLINE

The organization of the educational institution into terms such as quarters, semesters, or trimesters dictates to some extent the number of courses in the total curriculum and the content covered in each of the courses.

The curriculum outline usually lists the courses by terms indicating the class and shop or laboratory hours per week. The total of contact hours per week is usually included. Each course is normally given a code designation. Usually a determination of the number of credits allocated to each course is also included in the course outline. Other pertinent information may be provided as part of the course outline. Exhibits 7–1 and 7–2 illustrate typical course outlines.

COURSE DESCRIPTION

A description of each course is developed to convey to the reader a brief but concise statement of the content of the course. The course description should be written in language readily understood and meaningful to the reader. The title, code, credits, and prerequisites are usually given.

Scope and Purpose of the Course of Study

The course of study includes an outline of experiences, skills, knowledges, projects, demonstrations, and methods involved in teaching a particular subject. The objectives of the course, references, audiovisual aids, and instructional sheets are appropriately included in the course of study. The course of study really is a comprehensive plan of instruction which details the scope and teaching sequence of all of the activities of a particular course of the curriculum.

The purpose of the course of study is to provide an operating plan for the instructional program. Since the curriculum constitutes a master plan of all courses, so the course of study serves as an operating plan for the teacher for one course.

The course of study may be prepared by the teacher, by a curriculum

EXHIBIT 7–1

Mechanical Technology

		First Year	Hours Per Week		
Term 1			*Class*	*Lab*	*Credits*
MT	110	*Engineering Drawing	0	3	1
MT	130	Manufacturing Processes	2	2	3
AD	100	Orientation	1	0	0
LA	801	English	3	0	3
LA	810	Psychology	3	0	3
MA	140	†College Algebra and Trigonometry	4	0	4
PH	143	Physics	3	2	4
			16	7	18
Term 2					
MT	111	Eng. Drawing and Descriptive Geometry	1	3	2
MT	131	Manufacturing Processes	1	3	2
MT	155	Applied Mechanics	3	0	3
LA	802	English	3	0	3
MA	141	† Calculus with Analytic Geometry	3	0	3
PH	144	Physics	4	2	5
			15	8	18
Term 3					
MT	132	Manufacturing Processes	1	3	2
MT	156	Applied Mechanics	3	0	3
CH	104	‡Chemistry	3	2	4
LA	803	English	3	0	3
MA	142	†Calculus with Analytic Geometry	3	0	3
PH	145	Physics	3	2	4
			16	7	19
		Second Year			
Term 4					
MT	240	Precision Measurement	1	3	2
MT	257	Strength of Materials	3	3	4
MT	261	Fluid Mechanics	3	0	3
AD	120	Computer Programming	2	2	3
ET	127	Electricity	3	3	4
LA	830	Sociology	3	0	3
			15	11	19

* Students who have passed GS 132 Engineering Drawing with a C or better will not be required to take MT 110 Engineering Drawing.

† Students whose mathematics backgrounds are deemed adequate by the Admissions Office may start their mathematics sequence with MA 141. They would then take MA 142 and MA 240 in the second and third terms.

‡ Students who have passed a high school chemistry course with 80% or better or GS 140 with a C or better will not be required to take CH 104 Chemistry. Another course may be substituted by the department.

Source: Broome Community College, Binghamton, N.Y., 1971.

EXHIBIT 7–1 continued

Term 5			*Class*	*Lab*	*Credits*
MT	165	Metallurgy	3	3	4
MT	220	Mechanical Design	2	3	3
MT	260	Thermodynamics	3	3	4
MT	267	Statistical Quality Control	3	2	4
ET	128	Electricity	3	3	4
			14	14	19
Term 6					
MT	135	Materials and Processes	3	3	4
MT	221	Mechanical Design	3	3	4
MT	262	Thermodynamics	3	3	4
ET	129	Electronics	3	3	4
LA	820	Economics	3	0	3
			15	12	19

EXHIBIT 7–2

HEATING, REFRIGERATION, AND AIR CONDITIONING
TECHNOLOGY CURRICULUM

Course Number		*Course*	Contact Hours Per Week			
			Class	*Lab*	*Total*	*Credits*
		First Trimester				
TEC	102.68	Orientation and Applied Mathematics I	5	0	5	3
TEC	102.15	Blueprint Reading and Freehand Sketching	2	3	5	3
TEC	102.26	Elements of Electricity	2	3	5	3
TEC	102.01	Heating Systems	2	12	15	9
		Total	11	18	30	18
		Second Trimester				
TEC	102.69	Applied Mathematics II	5	0	5	3
TEC	102.16	Mechanical Drawing I	1	4	5	3
TEC	102.74	Applied Physics—Fluid Mechanics	2	3	5	3
TEC	102.02	Air Handling Devices	3	12	15	9
		Total	11	19	30	18
		Third Trimester				
TEC	102.22	Technical Communications I	5	0	5	3
TEC	102.19	Machine Shop Practice I	2	8	10	6
TEC	102.03	Refrigeration Systems	3	12	15	9
		Total	10	20	30	18

EXHIBIT 7–2 continued

Course Number		Course	Class	Lab	Total	Credits
		Fourth Trimester				
TEC	102.07	Temperature Control Systems ...	2	3	5	3
TEC	102.27	Elements of Electronics	2	3	5	3
TEC	102.62	Human Relations	5	0	5	3
TEC	102.04	Air Conditioning Systems	2	12	15	9
		(9 credits of electives may be approved) Total	11	18	30	18

Junior Diploma in Heating, Refrigeration, and Air Conditioning Technology Awarded Upon Successful Completion of the First Four Trimesters—72 Credits

Course Number		Course	Class	Lab	Total	Credits
		Fifth Trimester				
TEC	102.75	Applied Chemistry	2	3	5	3
TEC	102.08	Psychometrics and Load Calculations	2	3	5	3
TEC	102.09	Codes, Laws, and Ethics	5	0	5	3
TEC	102.05	Field Service Procedures	3	12	15	9
		Total	12	18	30	18
		Sixth Trimester				
TEC	102.64	Bookkeeping and Accounting	2	3	5	3
TEC	102.65	Principles of Salesmanship	5	0	5	3
TEC	102.31	Principles of Cost Estimation	5	0	5	3
TEC	102.06	Systems Design	3	12	15	9
		Total	15	15	30	18

Senior Diploma in Heating, Refrigeration, and Air Conditioning Technology Awarded Upon Successful Completion of Six Trimesters—108 Credits

Source: Technical Education Center, Clearwater, Florida.

specialist, or through a combination of efforts by many individuals. Some courses of study are developed and available from curriculum laboratories or units of local, state, and national government.

CONSTRUCTION OF THE COURSE OF STUDY

The course of study which is based upon occupational analysis is more likely to be valid and serve the purposes for which it is designed.

The formulation of the objectives for the course of study should occur early in the process of development. Expression of the objectives in terms of performance goals is essential if achievement of the goals is to be evaluated.

Included in the course of study should be that content identified to be essential for the course and which can be taught in the course. Eliminate from the course superfluous materials.

Arrange the content in harmony with established principles (derived from Thorndike's laws of learning and identified findings of occupational analysis). Most important of the principles of learning for this purpose are:

1. Readiness	it is easier to learn when the student is ready to learn.
2. Reinforcement	the more frequently we use what we have learned the more effective our performance becomes.
3. Value	the more useful the learning from the point of view of the learner the more willing he is to learn.
4. Doing	active involvement in the doing process is essential to effective learning in vocational education.
5. Known-to-unknown	the most effective order of learning is to begin with what the student already knows and progress to that which he does not know.
6. Success	nothing succeeds like success; help the student to be successful and to feel successful and he will usually desire to continue learning.
7. Confidence	confidence comes with success. Speed and accuracy grow out of ability to perform and confidence in that ability.
8. Challenge	provide continuing challenges and stimulating problems to hold the student's interest.

Plan methods of teaching the course to use the best techniques both from the point of view of work in industry and teaching techniques.

Outline the content, breaking the blocks of the course into main divisions and subdivisions in accordance with a three-point outline. Some prefer to separate the "doing" from the "knowing" activities thus developing two outlines instead of one.

Indicate the textbook or textbooks and references for the course. Books, handbooks, manuals, periodicals, and other biographical sources should be identified completely and correctly in accordance with an approved style manual.

A list of jobs and other activities planned for student involvement is an excellent addition to the regular outline of content of the course.

The grade level for which the course is intended should be specified as well as other information to provide complete details for those who may use the course of study but who did not have the opportunity to participate in its development.

Teaching aids such as films, film strips, slides, charts, etc., intended for use with the course should be listed in the course of study. The media for which the course is intended also should be identified. This is especially true if the course is designed for presentation by closed-circuit or broadcast television, video-tape, or the audio-video-tutorial systems of instruction.

An effective format should be selected for the course of study. The title, date, authors, and similar information usually appear on the title page. The format given here provides the main headings and the subheadings of a course of study.

SUGGESTED FORMAT FOR COURSE OF STUDY

I. Introduction
II. General considerations for the content
III. Objectives
    A. Broad behavioral objectives for the course of study
    B. Specific performance goals for each unit or lesson of the course
IV. Course description
V. Course outline
    A. Major divisions and subdivisions of each unit of instruction
    B. Shop or laboratory projects
    C. Instructional aids
    D. Texts and references
    E. Teaching methods and media
    F. Evaluation of performance
VI. Bibliography
VII. Other
    A. Suggested shop or laboratory layout
    B. Equipment list
        1. Specifications for ordering
        2. Cost estimate
    C. Supply list
        1. Specifications for ordering
        2. Cost estimate
    D. Unique feature to consider in planning facilities for the equipment required for the course

Summary

The word curriculum is used to express two concepts:

1. The sum total of all learning experiences provided in a controlled situation.
2. A group of courses selected to constitute the master plan of the educational program.

Validation of curriculum content and selection of topics of priorities are basic to effective curriculum construction. Eliminate content previously learned; also that content which can be postponed until the student is on the payroll job.

In planning curriculum:

1. Establish specifications.
2. Determine broad and specific objectives.
3. Gear to level of students.
4. Consider available space, equipment, personnel, and media.
5. Recognize entrance requirements.
6. Consider time to complete program.

Any controls which are imposed or inherent must be considered in curriculum building.

Methods of building curriculum include:

1. Analysis-based approach
2. "Scissors and paste pot" method
3. "Lifting" from other institutions
4. Combination of above

The analysis-based approach is most valid although a combination of all is very frequently used.

A curriculum outline should be provided to show:

1. Courses
2. Terms
3. Credits and/or hours per week

The course description used in the catalog and other literature should be brief, accurate, and descriptive.

The course of study is an instructor's broad operating plan for teaching a particular course. It is the basis for development of lesson plans, instructional information, and media.

Principles of learning relevant for curriculum builders include the concepts of:

1. Readiness
2. Reinforcement
3. Use

4. Involvement
5. Known-to-unknown
6. Success
7. Confidence
8. Challenge

QUESTIONS AND ACTIVITIES

1. Explain the differences between a curriculum and a course of study.
2. What is a validated curriculum?
3. Explain the meaning of a performance-based curriculum.
4. How are content priorities determined?
5. Discuss the problem of curriculum controls.
6. How is the philosophy of the curriculum builders related to and reflected in the curriculum?
7. What material should be included in the content of the curriculum?
8. Compare the function of designing a curriculum with that of designing a building.
9. Identify the major step in the process of curriculum building.
10. Which method of curriculum building is:
    A. Most valid?
    B. Fastest?
11. Select an occupation, preferably your own, and build the curriculum outline for it.
12. Why is a well designed course of study psychology-oriented?
13. What should be included in the course of study?
14. Explain the relative focus of the objectives of the curriculum as compared to the course of study.
15. Write the objectives of the curriculum of your occupation.
16. Write the objectives of a specific course of the above curriculum.
17. Build a course of study for one of the courses of the curriculum for your occupational field.

REFERENCES

1. Barlow, Melvin L., "Foundations for Vocational Education." Paper presented at the Citizens Conference on Man, Education, and Work, Colorado State University, October 3, 1968.

2. Emerson, Lynn A., *Technician Training Beyond the High School.* Raleigh: Vocational Materials Laboratory, State Department of Public Instruction, 1962.

3. Olivo, C. Thomas, "Analysis as the Basis for Effective Curriculum Development." Paper presented at the Institute for Curriculum Development Based on Occupational Analysis, Colorado State University, July 29, 1969.

4. Olivo, C. Thomas, "Curriculum Planning: The Basis for Program Development," *School Shop* XXIII (April, 1964), 43–45, 117.

5. *Vocational Education, The Bridge Between Man and His Work—Summary and Recommendations.* (Adapted from the General Report of the Advisory Council on Vocational Education.) Washington, D.C.: Superintendent of Documents, 1968.

Chapter 8 DEVELOPING PERFORMANCE GOALS

The purpose of all education and training is to modify the behavior of human beings. In some cases this modification consists of adding skills and knowledges, while in other cases it may involve changing habits or even attitudes.

The purposes of education are identified in the objectives, goals, or aims of education. Shades of meaning are reflected in the words selected. As used in this book the word "objective" denotes the broadest connotation of educational intents. Aims are more focused and used to refer to a small body of instructional materials or more sharply delineated educational intents. Goals are applied to units and lessons of instruction.

Performance goals are statements designed to reflect modifications or changes of behavior of the learner at the lesson or unit level. Performance goals are specific and explicit descriptions of education or training intents with particular emphasis on the actions performed on a payroll job. Behavioral objectives are applied to the curriculum in very broad terms only. Frequently behavioral objectives are used to imply performance goals. Both apply in about the same way to the intents of the instructional program but the difference lies in the breadth of focus.

The application of objectives, aims, and goals to various parts of the curriculum may be illustrated as in Figure 8–1.

Often the terminology is confused. Sometimes *broad* and *specific* objectives are the words used to describe the total range of meaning. In

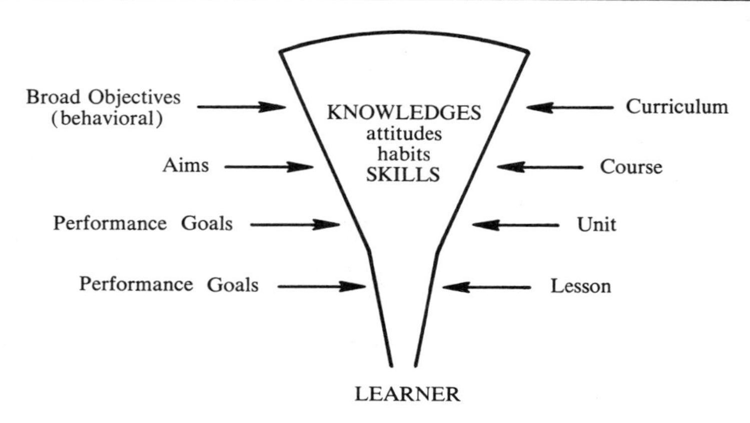

FIGURE 8–1

OBJECTIVES, AIMS, GOALS AS APPLIED TO CURRICULUMS

this approach *broad* objectives are the curriculum objectives and the course objectives. The *specific* objectives are those applied to units and lessons. Other persons have interpreted broad objectives as those applicable to the curriculum, and specific objectives as those designed for courses, units, and lessons.

Frequently in the literature the words objectives, aims, and goals are used interchangeably. Some authors use *goals* in describing the broadest intents. Sometimes the phrase *performance goals* is used to convey the same shade of meaning as "behavioral objectives."

The degree of specificity must be recognized as applicable from broad to very specific, in descending order to the curriculum, course, unit, and lesson, without regard to the word or term used to denote the meaning.

Performance Goals

During the 1960s a new emphasis was given to output by the development of a systems approach identified as PLANNING–PROGRAMMING–BUDGETING–SYSTEMS (PPBS). This approach was developed by the Rand Corporation and later introduced into various units of the federal government. The effect of the system was to change the emphasis from the input to the output. In other words the focus was placed on the results of the effort. This approach has meaning for vocational and technical education. Placing the emphasis on the results rather than the cost of education and training provides a much better approach to the student, parents, and the public.

Effective use of PPBS requires the establishing of valid performance goals. Often in the past the goals have been expressed in very broad, vague, and meaningless terms. These terms were expressed in language which could not be evaluated. The goals for objectives of instruction must be capable of being evaluated, otherwise it is impossible to know whether or not they have been achieved.

There are three main reasons for writing performance goals:

1. To inform the student what he is expected to learn in the training program
2. To acquaint the student with the conditions under which the learning is to take place
3. To provide the student with knowledge of the standards of achievement which are essential in the performance of the learning activities.

The success of any program of education depends upon the motivation of the student. Students must know what they are expected to do in order to be motivated. Performance results to a great extent from motivation. The performance goals must be so clearly established that there is no question in the mind of the student what he is expected to do. Curriculum materials must be developed to support the performance goals. The evaluation of the student's performance must flow directly from the performance goals established for the instructional program.

In other words, three elements are present in a performance goal. These are often described as conditions, description of task, and criteria of performance. The following are illustrations of performance goals:

1. Make a sketch of a straight shaft to be turned to a smooth finish to manufacturing tolerances with a diameter of 1.500 inches plus or minus 0.001″.
2. Adjust the complaint of a wholesaler who stated that he had received 8 percent of defective castings (Part Number 89CB17435). He said each casting failed at atmospheric temperature under normal dynamic loads. Returned castings have been submitted to the metallurgical laboratory for analysis and the report from the laboratory has been received. Using a personal interview and the laboratory report adjust the claim. Prepare appropriate memo and records of actions taken.
3. Explain how to solder a Number 10 wire to a terminal for standard current flow subject to a maximum pull of two pounds.

ADVANTAGES OF PERFORMANCE GOALS

Performance goals have many advantages both for the student and the instructor. The main advantage for both lies in the fact that each clearly knows what is the expected achievement upon the completion of the training program. The key words are *what* and *how*. The student will know if effective performance goals are developed. The setting of performance goals answers the question that so many students ask, "What must I be able to do and what must I know in order to complete this instructional program?" The following questions are related to "how." The question might be stated in terms of, "How well must I be able to do this task?" or "How many am I expected to complete in one hour?"

Success results from achievement of the performance goals. The old saying "Nothing succeeds like success" is true. The environment for learning is enhanced by letting students know what is expected of them—this means carefully developed performance goals.

Robert F. Mager demonstrated the importance of providing clear statements of the objectives in a laboratory-controlled situation. (4, p. 36) Students given clear objectives were able to perform the tasks with little instruction at a rate considerably faster than with either traditional instruction or programmed learning.

The advantages of clearly established performance goals to the student are:

1. Scope of learning tasks is positively identified within acceptable standards of achievement.
2. Time can be effectively used in seeking and achieving goals.
3. Resources available can be more effectively used to achieve goals.
4. Success is much more likely to result from the effort used.

The advantages to the teacher and the school are also significant. A great deal of time and careful analysis go into the writing of performance goals. Individuals attempting to write these goals must understand the basis for them and how to develop them.

To the vocational and technical educator performance goals have the following advantages:

1. The purpose of the instructional curriculum, course, unit, and lesson is much more clearly defined in terms of competencies essential for employment.
2. The time of the teacher can be used more effectively.
3. The resources available can be focused more effectively on the essential achievements.
4. The satisfaction of the concomitant learning with the teaching activities is greatly increased.

From the point of view of the educational institution the values of performance goals are also important since the results will be:

1. More effective learning and teaching.
2. Greater satisfaction on the part of employers of the graduates.
3. Fewer dropouts among the students.

4. Reduced costs of instruction.
5. An instructional program which can be more effectively evaluated. The evaluation then becomes a realistic basis for improving the instructional program.

CHARACTERISTICS OF PERFORMANCE GOALS

The basic concepts for performance goals are essentially the same as those for the writing of objectives in terms of behavioral changes resulting from instruction.

In the book, *Preparing Instructional Objectives,* Mager describes the process of developing behavioral objectives. (5) The process is fundamental to building realistic performance goals. In writing performance goals always be sure to:

1. Describe the results expected.
2. Tell what the learner is doing when correctly using the skills and knowledges needed on the job.
3. Be specific rather than general.
4. Use clear and simple language.
5. State the standards required on the job for successful performance. This usually requires a statement of quality, quantity, or both.

ELEMENTS OF PERFORMANCE GOALS

Study and analysis emphasize the three main elements of performance goals. These are repeated and discussed in greater detail below. To be complete each performance goal should state:

1. The *conditions* under which the task is performed.
2. The *description* of the task.
3. The *criteria* for recognizing performance.

Conditions. The statement of a performance goal contains a description of the conditions under which the performance will take place. All important conditions are identified which may be applicable to the job or task. The description of the conditions is in specific terms so that the performance of the student can be accurately compared against the required standards of the job in industry, business, or agriculture.

Description of Task. The second part of a performance goal defines the task. The definition needs to contain exact descriptions of the task in terms of quality, quantity, or time elements of task performance.

Criteria of Performance. The third part of a performance goal states the criteria of performance so that the achievement of the goal can be recognized. The steps are identified and listed which are necessary to performance of the job or task. Key points may be identified for each of the steps if desired.

KEY WORDS IN WRITING PERFORMANCE GOALS

Writing performance goals is simplified by keeping in mind the approach used by the *Dictionary of Occupational Titles.* (3) The premise used in Volume 2 of the *Dictionary* is that every job requires a worker to function in relation to data, people, and things, in varying degrees. Modifications of these variations of meaning if the exact words of the code are not appropriate will often provide an aid for determining the words to select in writing performance goals. The numbers at the left refer to a general hierarchical system used in the code of the *Dictionary.* The number of the digit of the code is given with the category. (See Exhibit 8–1.)

EXHIBIT 8–1

DICTIONARY OF OCCUPATIONAL TITLES
CLASSIFICATION PLAN

DATA (4th digit)	PEOPLE (5th digit)	THINGS (6th digit)
0 Synthesizing	0 Mentoring	0 Setting-up
1 Coordinating	1 Negotiating	1 Precision working
2 Analyzing	2 Instructing	2 Operating—Controlling
3 Compiling	3 Supervising	3 Driving—Operating
4 Computing	4 Diverting	4 Manipulating
5 Copying	5 Persuading	5 Tending
6 Comparing	6 Speaking—Signaling	6 Feeding—Offbearing
	7 Serving	7 Handling

Included in the materials developed as part of the Vocational Instruction System (VIS) of the State of Washington (6) are the selected key words shown in Exhibit 8–2.

SPECIFYING PERFORMANCE GOALS

Performance goals must be neither too general nor too specific. The teacher and the student must know exactly what is to be learned and how satisfactory achievement is to be demonstrated. The performance goals must be identified in terms of observable and measurable performance. Action verbs are needed to describe the performance desired. The conditions under which the student will perform the action must also be described. Realistic performance goals must be based upon student performance under joblike conditions; otherwise validity will be lacking. The standards of the performance goals usually consist of ac-

EXHIBIT 8–2

AIDS OR KEY WORDS IN WRITING OBJECTIVES
(PERFORMANCE GOALS)

Must be able to	Calculate	Construct	Protect
Will	Present	Draw	Service
Should	Manipulate	Analyze	Visual check
Demonstrate	Illustrate	Direct	Inspect
Perform	Recognize	Recommend	Clean
Examine	Identify	Modify	Purge
Correct	Take action	Revise	Decontaminate
Actuate	Remedy	Test	Composite test
Define	Write	Checkout	System align
List	Compare	Calibrate	Store
Explain	Differentiate	Remove	Handle
Show	Locate	Install	Monitor
Relate	Design	Replace	Operate
Can	Determine	Repair	System installation
	Without aid of	Overhaul	Subsystem assembly
	Subsystem disassembly		

curacy and time. Again the standards should be similar to those found on the payroll job. Some samples of performance goals are:

1. The trainee will match the name of the electronic components with the symbols on a schematic drawing of any radio circuit. All symbols must be correctly identified.
2. The trainee will list in order all the major moving parts in the power train of a standard shift car. The list will be produced from memory with no assistance. The trainee may omit, have out of sequence, or misname no more than one major part.
3. The trainee will start a mobile motor generator, type A–26, using the operating manual as a guide. He must have the generator operating within the specified tolerances (117 ± 3 volts AC and 60 ± 3 cycles) within three minutes.
4. The trainee will perform simple addition on a calculator, Smith Model 5–A. He must set up the machine and carry out the addition of sets of five-digit numbers with decimals to three places without the instruction booklet. Ten problems, each with ten five-digit numbers, must be added within five minutes with no procedural or arithmetic errors.
5. The trainee will locate and identify malfunctions in the electrical system of a car. The malfunctions will be shorts and opens induced at logical locations in the system. The trainee may use screwdriver, pliers, multimeter, schematic drawings, and the repair manual. Eight of ten of the malfunctions must be located and identified correctly within fifteen minutes. (1, pp. 73–74)

HIERARCHY OF INTENTS AND PERFORMANCE GOALS

The general objectives closely follow the occupational or job analysis and are used as the broad descriptive statements of the curriculum. Specific performance goals must be written at the level of the operation (task) or job if it is to be most useful. At this level the basic skills and knowledges must be learned; therefore, the performance goals must be focused at this point for maximum clarity, meaning, and evaluation of performance.

Performance goals of a somewhat broader nature can be written at the level of the unit. The degree of specificity will not be as refined as if the point of focus is at the operation level. The amount of time available for development of performance goals, curriculum materials, and performance evaluation may influence the level at which the performance goals will finally be written.

A comparison can easily be made between the level at which performance goals are written and instruments used to measure a linear dimension. (See Exhibit 8–3.)

EXHIBIT 8–3

COMPARISON OF WRITING PERFORMANCE GOALS
TO USING INSTRUMENTS FOR LINEAR MEASUREMENTS

Linear measurement of shaft with a:	Developed behavioral objective or performance goal written at the level of:
Yardstick in increments of 1 inch	Curriculum—behavioral objectives
Rule in increments of 1/64 of an inch	Course—behavioral objectives
Micrometer in increments of .001 of an inch	Unit—performance goals
Micrometer in increments of .0001 of an inch	Job or operation (task)—performance goals

ORDER OF PERFORMANCE GOALS

Building performance goals into the instructional system in the most meaningful way requires proper sequencing. The order of performance goals should follow a plan of functional context sequencing. (2, p. 167) Inherent in such functional context sequencing are concepts such as the following:

1. Provide the student with the "big picture."
2. Make each part relevant for the student and relate logically to the total sequence.

3. Teach the "whole" first and then teach each of the parts.
4. Begin with simple jobs and advance to more complex ones.
5. Add reinforcement as needed.
6. Provide essential repetition to build skill and confidence.
7. Introduce new skills and knowledges in succeeding operations.
8. Select jobs which are most essential for the achievement of the identified goals.

SIGNIFICANCE OF PERFORMANCE GOALS

If carefully developed performance goals are constructed, the instructional process will be more meaningful to the student. Both the student and the teacher will be able to correctly determine whether or not the purposes of instruction have been achieved. The performance goals will in themselves be a power aid to learning for the student. He will know what is expected of him.

Illustrations of Objectives and Performance Goals

The following were developed by students in teacher-training classes. The objectives were developed in group sessions during the study of behavioral objectives and performance goals as illustrations of curriculum course and lesson objectives.

A. CURRICULUM OBJECTIVES

Students will, upon completion of the curriculum:

1. Be capable of performing entry-level clinical laboratory procedures in hospitals, clinics, and medical research centers. (medical lab technician)
2. Demonstrate good work habits, punctuality, and pride in their work. (drafting)
3. Develop skills required to isolate malfunctions in computer equipment. (computer repair technician)
4. Demonstrate good safety habits around machine shop equipment and fellow students. (machine shop)
5. Be capable of explaining and demonstrating common welding processes. (welding)
6. Develop skills in the operation and management of farm imple ment dealer service departments. (farm mechanics)
7. Disassemble, inspect, and reassemble components using required special tools, precision measuring devices, and manufacturer's service information and procedures. (automotive)

8. Operate, maintain, calibrate, and analyze electronic equipment. (electronic technician)
9. Develop skills required to isolate malfunctions in electronic equipment. (electronic technician)
10. Repair microelectronic circuits in accordance with NASA standards. (microelectronics repair technician)
11. Perform dental office procedures efficiently. (dental)
12. Perform academically and clinically the requirements necessary for national certification in dental assistance. (dental)
13. Be prepared for successful employment at entry-level positions which support wildlife professionals. (wildlife management)
14. Have practical working knowledge of fundamental principles of wildlife management which will enable them to develop competence necessary for continued gainful employment. (wildlife management)
15. Develop a philosophy toward life that will enable them to be productive members in the welding trades. (welding)
16. Develop ability to perform basic principles of architectural graphics. (architecture)
17. Perform basic skills in nursing arts used in patient care. (health)
18. Enter and function with basic skills necessary for employment in auto mechanics. (auto mechanics)
19. Work safely and effectively in a machine shop. (machine shop)
20. Develop skill, knowledge, and techniques required to perform advanced precision measurements in electronics. (electronics)
21. Be able to identify, analyze, and repair diesel engines. (diesel engines)
22. Inspect, clean, maintain, and operate APS 108 radar set. (navigation system)
23. After receiving a two-year high school certificate in vocational automotive mechanics, be able to continue, if they so desire, their education in an advanced automotive institution. (auto mechanic)
24. Be prepared for entry-level employment as an automotive maintenance technician. (auto mechanic)
25. Be able to prepare manuscripts on technical subjects for submission to publisher. (tech writing)
26. Be able to produce, in mailable typewritten form, most of the rough draft material handed to her. (general clerical worker)
27. Be able to set up and operate a sheet metal press. (sheet metal development)

28. Meet entry-level requirements for employment as an industrial maintenance technician. (industrial maintenance)
29. Be prepared for an entry-level job as a data systems manager. (data systems supervisor)
30. Be prepared for entry-level employment as a test equipment calibration technician. (electronics)
31. Be a competent entry-level ______ technician. (all fields)
32. Be able to obtain entry-level employment as a ______ technician. (all fields)
33. Be an active, productive, and well-informed member of society. (all fields)
34. Upon completion of this curriculum, be able to communicate directly with management, scientists, and production personnel of his specialized area. (could be professionals of other fields)
35. Be capable of growing into positions of increasing responsibilities. (all fields)
36. Have a wide background in the diverse areas of the ______ field. (all fields)
37. Be capable of using and/or operating the tools and equipment of the profession. (all fields)
38. Be capable of maintaining the tools and equipment of the profession or occupation. (all fields)
39. Using specifications, drawings, layouts, blueprints, schematics, and other such communications media, be able to communicate with other personnel in the line and/or the customers. (electronics)
40. Have a sustaining sense of responsibility toward nursing by keeping current in the knowledges and skills related to administration of health care. (health occupations)

B. COURSE OBJECTIVES (AIMS)

Student will, upon completion of the course:

1. Be proficient (determined by known standards) in preparing, testing, and analyzing blood specimens for hematological examinations. (medical lab technician)
2. Demonstrate the correct use of all eleven scales of the architect scales. (drafting)
3. Be able to troubleshoot and repair the Burroughs D-84 Computer. (computer repair technician)
4. Compute the proper engine lathe speeds and feeds and types of cutting tool for cutting various diameters of various materials. (machine shop)

5. Test for leaks in an oxygen-acetylene manifold system. (welding)
6. Develop an understanding of the process involved in transforming fuel to power in an internal combustion engine. (farm mechanics)
7. Disassemble, inspect, and reassemble a standard transmission using required special tools, precision measuring devices, and manufacturer's service information and procedures. (automotive)
8. Determine the symbols used to identify electronic circuit components according to military specifications. (electronic technician)
9. Apply mathematics in solving electronic problems. (electronic technician)
10. Be able to use resistive soldering devices. (microelectronics repair technician)
11. Perform calibration on specified pieces of Hewlitt-Packard equipment to meet manufacturer's specifications within certain time limits. (electronics)
12. Demonstrate use of cutting tools in machine shop. (machine shop)
13. Administer accurately and quickly medications to patients according to doctor's orders. (health)
14. Select type of dental stone used for study models and measure water and powder in correct ratio, manipulate, and pour study model. (health)
15. Demonstrate proper technique in lighting an oxyacetylene torch. (welding)
16. Identify as to species, sex, and age selected species of water fowl, upland birds, and mammals. (wildlife)
17. Calculate range of errors in measurements of power and attenuation to determine degree of accuracy. (electronics)
18. Remove and replace automotive power train in accordance with manufacturer's specifications and time. (auto mechanics)
19. Use techniques and mapping instruments to construct a field map of specified land area which can be used effectively by other persons. (wildlife)
20. Give complete immediate postoperative patient care in recovery room. (health)
21. Measure precision parts within .0025 inch with a micrometer. (diesel precision measuring)
22. Be able to illustrate radar presentation representative of all five

modes of computer operation. (radar data–presentation systems)

23. Be able to disassemble a two-section blanking and punching die in 20 minutes. This does not include the guide pins and guide pin bushings. (sheet metal development)
24. With the proper test equipment, hand tools, and motor flat rate manual, be able to overhaul and make adjustments on any American made carburetor. (auto-carburetion)
25. Be able to locate, bore, and counterbore holes to an accuracy of .0002 inch with the Pratt and Whitney 2-E jig boring machine. (machine shop)
26. Be able to prepare rough drafts of illustrated material in detail so that artists may complete accurate drawings *without* further directions. (tech writing–production)
27. Given the proper equipment and a prepared transcribing tape, be able to type mailable letters directly from the tape. (general clerical worker)
28. Remove and replace a compressor on an American Standard refrigeration unit. (refrigeration compressors)
29. Be able to operate the APS-108 radar system and perform the transmitter control frequency adjustment in accord with a technical order. (bomb, nav systems RDPS)
30. Be able to check all hydraulic pressure points and diagnose all probable trouble points in the pump relief valve circuit. (hydraulics)
31. Solve data processing problems using the five steps of problem solving. (data processing)
32. Be able to test the equipment for defects using electrical measuring instruments. (electronics)
33. Without the aid of references, be able to identify thirty chronic diseases and the body systems they primarily affect. (practical nursing)
34. Interpret manufacturer's specifications related to technical manuals, drawings, layouts, blueprints, and schematic diagrams. (engineering technology)
35. Analyze the nature and operation of computer circuits he will encounter throughout the remainder of this curriculum. (electronics)
36. Have a working knowledge of basic arithmetic, algebra, and introductory trigonometry to grasp the electrical concepts taught in concurrent and subsequent courses in this curriculum. (electronics)

37. Be able to compute charges for labor and materials, using charts, checklists, manuals, directories, and calculators. (business and distributive education)
38. Be able to develop technical skills essential to the administration of basic patient care. (nursing)
39. Interpret and/or follow directives and/or instructions in written, oral, diagrammatic, or schedule form. (all fields)
40. Identify and demonstrate the use of the Binary Coded Decimal symbols. (data processing)

C. LESSON OBJECTIVES (PERFORMANCE GOALS)

The student will, upon completion of the lesson:

1. Be able to calculate and perform a white blood count with 95 percent accuracy (standard) within fifteen minutes. (medical lab technician)
2. Given that self-test 14 is GO and self-test 15 is NO GO, write and execute a test sequence to determine the operational status of module 3Z2 of the D-84 computer in fifteen minutes. (data processing)
3. Draw a three-dimensional object in one plane to half scale within 1/32″ accuracy. (drafting)
4. Set up and perform a straight turning operation on mild steel using an engine lathe with ±.005″ tolerance. (machine shop)
5. Test for leaks in an oxygen-acetylene manifold system using leak test solution. (welding)
6. Given a four cylinder in-line gasoline engine perform a compression test and interpret the readings within twenty minutes. (farm machines)
7. Disassemble a standard three-speed synchromesh transmission in accordance with manufacturer's procedure using the proper tools. (automotive)
8. Interpret color bands on fifteen carbon resistors in fifteen minutes. (electronic technician)
9. Calculate the resonant frequency of a bridged T oscillator to within ±5 hertz. (electronic technician)
10. Using a resistive soldering device connect and solder a lead to a connector pin within ten minutes in accordance with NASA Handbook SP5002. (microelectronic repair technician)
11. Machine a number two Morse taper on a lathe to ±.002″ on large diameter. (machine shop)
12. Make an open surgical bed in five minutes. (health)
13. Pass a Levine tube into a patient's stomach. (health)

14. Using a human skull, locate and name all the bones of the oral cavity. (health)
15. Adjust an oxyacetylene welding flame to be used for bronze welding as demonstrated by the instructor. (welding)
16. Measure a three dB in-line attenuator to within ±.001 dB of instructor's reading, taking no more than fifteen minutes. (electronics)
17. Band water fowl and list the date on a U.S.D.A. form #201. (wildlife)
18. List the common emergencies that occur in the dental office and their symptoms and treatment. (health)
19. Assemble a given set of eight pistons to their individual rods in twenty-five minutes according to manufacturer's specifications. (auto mechanics)
20. In one hour draw a freehand sketch of a stairway in pencil according to instructor's demonstration. (architecture)
21. Calibrate an HP 410B vacuum tube voltmeter with the KO-1 voltmeter calibration system to an accuracy of 3 percent within two hours. (voltmeter calibration test equipment repair)
22. Using the correct control panel, wire the reproducer to accomplish field selection and compare. One hundred percent accuracy is required. (machine operator selective reproduction)
23. Remove transistor 01 from the J1 PC board without damaging the transistor of PC board. (soldering techniques)
24. Correctly assemble a Fisher WCV 25 pneumatic valve within thirty minutes. (pneumatic valves)
25. Remove, solder, test, and replace the fuel tank in a late model automobile. (automotive fuel tanks)
26. Given filled-out time cards and tax tables, compute net pay for individual employees. (payroll)
27. Given a rough draft copy, type at least four letters and envelopes in one hour with three or less corrected errors, AMS style. (typing)
28. Adjust cracking pressure to factory specs on a Rossemaster throw-away fuel injector in ten minutes. (diesel adjusting fuel injectors)
29. In each group, correctly prepare ten curriculum, ten course, and ten lesson objectives within an hour and one-half. (technical writing)
30. Given samples of hot rolled steel, cold rolled steel, aluminum, and stainless steel, list each one by surface characteristics. (industrial materials)

31. Be able to operate the card-reader without supervision or use of reference materials. (data processing)
32. Be able to make an unoccupied bed, free of wrinkles, within seven minutes. (nursing)
33. Be able to calibrate a vacuum tube voltmeter to manufacturer's specifications in forty minutes or less. (electronics)
34. Without the use of reference materials, be able to identify and demonstrate the use of each flow charting symbol, achieving 95 percent accuracy. (data processing)
35. Be able to solder tube socket terminal connections within two minutes to NASA standards. (electronics)
36. Be able to change an abdominal dressing (patient selected by the instructor), using sterile techniques, within fifteen minutes. (health occupations)
37. Given a list of twenty vital signs, be able graphically to record them with 100 percent accuracy. (health occupations)
38. Within ten minutes, align a five-tube superheterodyne broadcast receiver from the production line using only a VTVM and signal generator. (electronics)
39. Be able to adjust the distributor points of a Ford within five minutes, using the sun reflectronic analyzer. (automotive)
40. Be able to set up and remove magnetic tapes from the magnetic tape unit without supervision. (data processing)

Summary

Intents of education are indicated by such descriptive words as objectives, aims, and goals.

While shades of meaning are reflected in the words selected (objectives, aims, goals), these words may not be used in precisely the same way by different individuals.

In planning the intents of education the broadest objectives should be applied to the curriculum. These are behavioral objectives. The focus is sharpened when planning the intents of the course. These are the aims of the course. As the intents of each unit and lesson are delineated the focus of purpose becomes very specific. These very specific purposes of learning and teaching are called performance goals.

Performance goals should:

1. Inform the students specifically of the purpose of the unit or lesson.
2. Identify conditions under which the learning will take place.
3. Establish the standards of performance.

Well-planned performance goals save both the instructor's and the students':

1. Time
2. Effort
3. Resources

Performance goals also:

1. Improve quality of performance.
2. Lead to feelings of success.

Three basic elements of performance goals are the:

1. *Conditions* under which the task is performed.
2. *Description* of the task.
3. *Criteria* for recognition of performance.

Selection of a specific, sharply descriptive word of limited meaning is the *key* to writing performance goals.

If descriptive *behavioral objectives* are constructed for the curriculum and focused *performance goals* are formulated for the units and lessons, they will enhance, expedite, and improve both learning and teaching.

QUESTIONS AND ACTIVITIES

1. Distinguish between the meaning of objectives, aims, and goals.
2. What is the difference between a behavioral objective and performance goal?
3. Why should behavioral objectives be developed for a curriculum?
4. Why should performance goals be formulated for a unit or lesson?
5. What is PPBS?
6. Contrast the advantages with the disadvantages of using performance goals in the instructional process.
7. How does Mager describe the significance and process of development of behavioral objectives?
8. What are the essential elements of a performance goal?
9. Explain the basic concepts and features of performance goals.
10. Add several "key words" of your own that may be used in writing performance goals.
11. Diagram the hierarchy of intents as used with program planning from the curriculum to the lesson.
12. What did Mager mean when he said, "If you give each learner a copy of your objectives, you may not have to do much else?"

13. Build the objectives for your curriculum in terms of behavioral objectives.
14. Construct appropriate educational intents for one or more courses of your curriculum.
15. Formulate suitable performance goals for one or more units of a course.
16. Plan and write focused performance goals for several lessons of a unit.
17. Illustrate by examples in your own occupational field the differences in shade of intent (specificity) needed in planning the curriculum, course, unit, and lesson.

REFERENCES

1. Butler, F. Coit, Jr., *Job Corps Instructional Systems Development Manual.* Washington, D.C.: U.S. Office of Economic Opportunity, January 9, 1967.
2. Committee for Economic Development, *The Schools and the Challenge of Innovation.* New York: McGraw-Hill Book Company, 1969.
3. *Dictionary of Occupational Titles,* Volume II. Washington, D.C.: U.S. Department of Labor, 1965.
4. Mager, Robert F., "The Need to State Our Educational Intents," *Technology and Innovation in Education* (Prepared by the Aerospace Education Foundation). New York: Frederick A. Praeger, 1968.
5. Mager, Robert F., *Preparing Instructional Objectives.* Palo Alto: Fearon Publishers, 1962.
6. Wimer, Frank H., "State of Washington—Vocational Instruction System (VIS)." A paper presented at the Institute on Occupational Analysis as a Basis for Curriculum Development at Colorado State University, July 28–August 1, 1969. Olympia: State Department of Education, 1969.

Chapter 9 UNITS OF INSTRUCTION

The learning process is expedited through organization of the curriculum. Meaningful relationship of the parts of the curriculum to each other and to the total world of work adds the needed relevance to the experiences for the student.

Psychology of the Learning Process

Application of psychology to curriculum organization increases the rate of learning and makes learning easier and more meaningful to the student. Some of the most significant concepts of learning for curriculum development are the following:

1. Most, if not all, learning takes place when we want to learn something.
2. Most, if not all, learning is a matter of finding out what things go together—what things belong together.
3. An important thing in learning is remembering, and we remember things best which have:

a. made a deep impression on us, or
b. which we have used or practiced a great deal.

4. Practice makes learning stay with us.
5. Learning is easier and more likely to take place if the things we are learning fit into some whole picture or act.
6. Words and symbols are important tools of learning because they stand for ideas and things. We tie facts, events, and experiences to words.
7. There are certain ways of going about learning that are good, and the student needs to form habits of learning which can help him.
8. Learning can only take place through experiences—experiences such as seeing things, handling them, working with them, and hearing about them. (1, p. 89–90)

Learning is easier and teaching is more gratifying when success is evident. Success is more likely to result when the components of learning fit together to provide a complete picture. Most learning then is a matter of finding out what parts can be effectively combined to make larger and more meaningful experiences.

Recap of Derivation of Content for Units

Validated instructional content is a product of analysis. Analysis of the payroll job or occupation can be combined with performance on the job as recommended by the advisory committee and exploration of reliable secondary sources such as handbooks, manuals, textbooks, and reference books to form the base for building curriculum.

Competencies needed for entry-level jobs should be translated into performance goals and into a curriculum. The curriculum, in most cases, should be divided into courses to fit the organizational pattern of the educational institution. Each course in turn may be composed of several units. Units will have *doing* and *knowing* emphasis in combination. Some units are identified more with related or applied instruction, while other units are primarily designated to provide instruction in shop or laboratory activities. The learning process is enhanced by interrelationships between *doing* and *knowing* units. Integration of concepts with applications is essential for meaningful vocational education. Students really have not learned until they can apply in a realistic sense the principles or theories to a problem that needs solution.

Units Defined

A unit has been defined as a group of similar or related objectives that make up a whole. In the vocational curriculum a unit of instruction is a combination of operations, jobs, and related or applied information that focuses on a major topic.

In a unit the activities in the shop or laboratory will require the same kinds of skills and knowledges applied in varying degrees. Similar tools and machines will be used in completing the operations or jobs involved.

The content in a unit specifically identified with a course of related or applied instruction will focus on a major topic of importance to the course. Each of the lessons of the unit will develop a certain part of the knowledge base required for mastery of the topics being studied.

Integration of shop or laboratory activities with related or applied class work is vital to the success of both the shop class and the related class. Without such integration the related classes will have little meaning and the student will be ill prepared to proceed with the work to be done in the shops.

A unit of work is a complete, coherent learning activity involving an integration of industrial processes or scientific principles with knowledge, skills, and attitudes essential to successful application on a payroll job.

Carter V. Good has defined a unit of instruction as:

> A major subdivision of a course of study, a textbook, or a subject field, particularly a subdivision in the social studies, practical arts, or science; an organization of various activities, experiences, and types of learning around a central problem, or purpose, developed cooperatively by a group of pupils under teacher leadership; involves planning.*

While Good's definition is broad and includes applications to all fields of education, it also covers vocational education. In some situations the students participate in the planning of the unit of instruction. This is more likely to occur in relation to introductory programs of vocational education than in advanced instructional programs whose objectives are sharply focused on identified jobs in industry.

UNIT IN RELATIONSHIP TO CURRICULUM

The relationship of the unit to the total curriculum, the course of study, and the instructional sheets is shown in Figure 9–1. The unit is

---

* From *Dictionary of Education,* 2nd Edition, edited by Carter V. Good, p. 587. Copyright 1959 by the McGraw-Hill Book Company. Used with permission of McGraw-Hill Book Company.

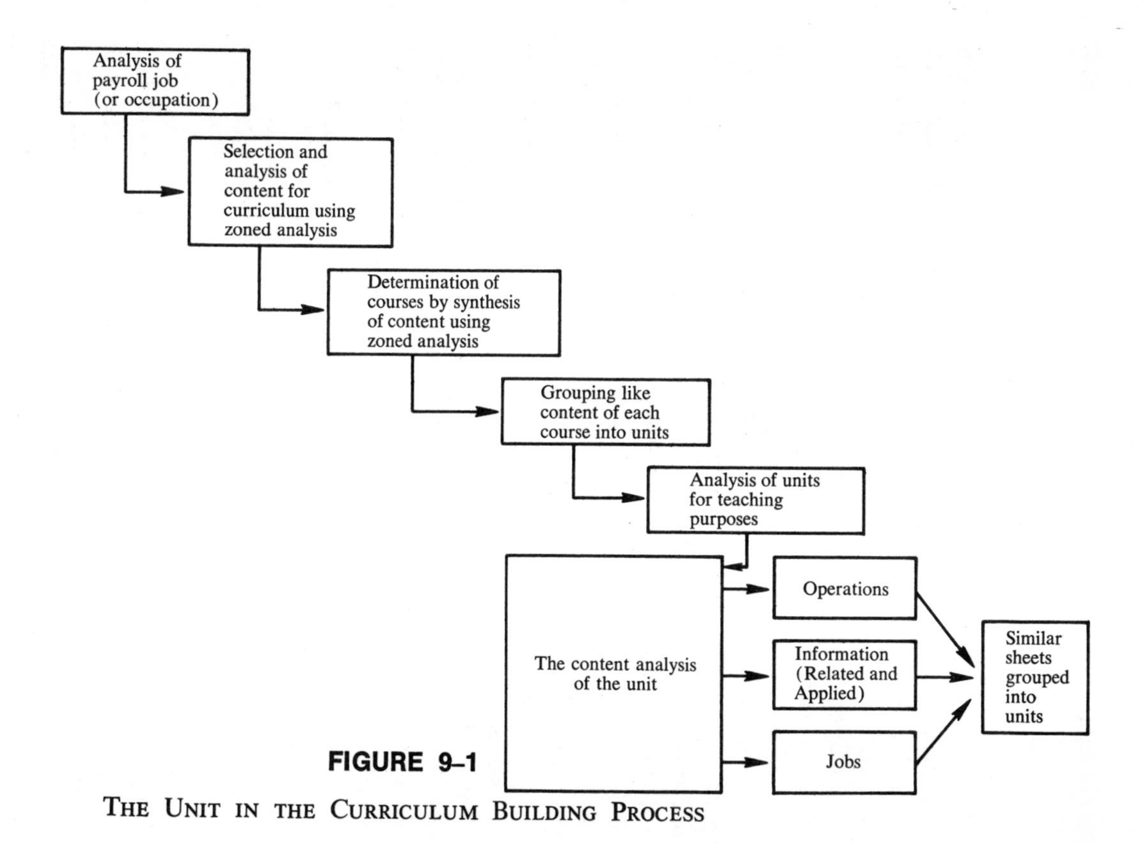

FIGURE 9–1

THE UNIT IN THE CURRICULUM BUILDING PROCESS

obtained by dividing the courses of study into compatible teaching increments. Each unit may be composed of several operation sheets, job sheets, and information sheets as well as other kinds of lesson materials.

The size of the unit should be determined in accordance with the nature of the content and the span of interest of the level of students in the class. The kinds of jobs and the complexity of the content will influence the length of the unit. The unit of instruction should be readily adaptable to different levels of students. The background of the students will affect the length required to achieve acceptable performance. Other factors to consider in building units of instruction are:

1. Motivation of the students
2. Availability of learning materials
3. Number of work stations
4. Degree of proficiency required

UNITS OF "DOING" AND "KNOWING"

Analysis provides information about what individuals must be able to do on the payroll job. This may be called the *doing* unit. The information is also derived from the analysis as to what the person must know who is to perform the job satisfactorily. This is sometimes referred to as *knowing* units.

The *doing* elements become the basis for the instructional program in the shop or laboratory, while the *knowing* elements are integrated with other vital or *nice-to-know* information in related or applied classes of instruction.

The units can easily be identified as divisions of a section in a course of study. The unit topic may not distinguish between the *knowing* and the *doing* in the outline as both may be listed in a combined treatment.

Building the Unit into Instructional Materials

When the decision has been made relative to the units to be taught, it is then necessary to develop the related instruction sheets to be used with the unit. The performance goals for the unit must be clearly identified and stated first. Then the procedure consists of:

1. Constructing the content analysis
2. Developing the operation sheets
3. Building the job sheets
4. Formulating the information sheets
5. Determining the content for the assignment sheets

6. Identifying other instruction content, materials, and media to be used
7. Assessing learning in conformance with the identified performance goals

ILLUSTRATION OF UNITS

Exhibit 9–1 illustrates the initial course outline which is the result of the analysis of the occupation into large blocks of learning (Sections) and the initial breakdown of the sections into units for several specific occupational training programs.

EXHIBIT 9–1

SAMPLE COURSE OUTLINES

EXAMPLE I. ELECTRONIC MECHANIC (ANY INDUSTRY) (ENTRY)

- Section I. Fundamentals of Electricity
  - Unit 1. Introductory Material
  - Unit 2. Basic Electricity
  - Unit 3. Current Electricity
  - Unit 4. Generation and Application of Electricity
  - Unit 5. Electric Motors
- Section II. Electronics
  - Unit 1. The Electron Tube
  - Unit 2. Semi-conductors
  - Unit 3. Practical Application of the Electron Tube and Semi-conductors
- Section III. Industrial Application

EXAMPLE II. WELDER COMBINATION

- Section I. Introduction
  - Unit 1. Orientation
  - Unit 2. Introduction to Materials and Equipment
- Section II. Welding
  - Unit 1. Oxyacetylene Welding
  - Unit 2. Grinding
  - Unit 3. Electric Arc Welding
  - Unit 4. TIG Welding
  - Unit 5. MIG Welding
  - Unit 6. Pipe Welding
  - Unit 7. Cutting
  - Unit 8. Additional Related Theory
- Section III. Drawings
  - Unit 1. Blueprints
  - Unit 2. Abbreviations and Symbols

EXHIBIT 9–1 continued

EXAMPLE III. AUTOMOBILE MECHANIC (ENTRY)

Section I. Introduction
Unit 1. Orientation
Unit 2. Safety
Unit 3. Survey of Shop and Shop Equipment

Section II. Tools, Toolroom, and Miscellaneous Parts
Unit 1. Hand, Bench, and Floor Tools
Unit 2. Miscellaneous Parts Used on Vehicles
Unit 3. The Toolroom

Section III. The Automobile Chassis
Unit 1. The Frame
Unit 2. Suspension Systems
Unit 3. Steering Systems

Section IV. Automotive Electricity and Electric Service
Unit 1. Fundamentals of Electricity
Unit 2. Lighting Circuits
Unit 3. Other Circuits of the Automobile

Section V. The Engine and Associated Systems
Unit 1. The Power Plant and Associated Systems
Unit 2. Problems, Service, and Repair of Cylinder Block Assembly
Unit 3. Service and Repair of Power Plant Units (Accessories to the Engine)
Unit 4. Lubrication of the Power Plant

Section VI. The Power Train (Drive Train)
Unit 1. The Clutch and Clutch Housing
Unit 2. Transmissions
Unit 3. The Drive Line (Universal Joints and Propeller Shaft)
Unit 4. Rear Axles and Differential

Section VII. Wheels, Tires, and Brakes
Unit 1. The Brake System
Unit 2. Tires and Wheels
Unit 3. Wheel Alignment

Section VIII. The Automotive Body
Unit 1. Doors, Windows, and Seats
Unit 2. Hood, Back Deck, and Ventilators

Section IX. Automobile Accessories
Unit 1. Air Conditioning and Heaters
Unit 2. Windshield Washers and Wipers, and Other Accessories

Section X. General Lubrication

Source: A Guide for Building a Course of Study. Nashville: Tennessee State Board for Vocational Education, pp. 17–19.

These examples demonstrate the analysis of three different occupations broken down into the major learning blocks, and in addition show how each large block has been further divided into units.

ADVANTAGES OF THE UNIT METHOD

The unit method aids learning and teaching because:

1. It grows out of analysis of the curriculum and the payroll job.
2. It integrates the *how* with the *what* and the *why* to provide meaningful experiences to the student.
3. It provides better organization and more comprehensive treatment of the related skills, knowledges, attitudes, and habits to be mastered.
4. It provides for orderly and organized progression through the course of study.
5. Reinforcement is in terms of relevant related experiences.

Summary

The organization of learning experiences and teaching content into units of instruction is sound, both logically and psychologically. Where several members of the faculty teach content as a team, joint planning of the content of the units is desirable.

If the unit approach is not used, students frequently experience difficulty in relating the fragments of learning gleaned from the various classes into a meaningful and relevant experience. Even with the unit method, the instructors must constantly seek methods of tying the content of the various sections and courses together so the students may better apply each portion to the total treatment required for entry into employment.

The unit is a plan for grouping closely related instructional elements and teaching these consecutively over a period of a few days or weeks.

Operation sheets, information sheets, and assignment sheets should grow out of units of instruction. However, some instructors have taken the approach of grouping operation sheets together with information sheets and assignment sheets as well as other similar related instructional elements to build a unit. The same approach may be used in the grouping of jobs. Really, this is another way of achieving the same result as will occur by grouping related performance goals under broad behavioral objectives.

QUESTIONS AND ACTIVITIES

1. What is meant by a unit of instruction?
2. Why should instructional content and processes be organized into units?
3. Explain the implications of psychology for the unit plan.
4. How is the content for units determined?
5. Discuss the relative merits of grouping performance goals vs. grouping instructional sheets.
6. What factors determine the size of the unit? Explain.
7. Explain the steps in the process of building a unit of instruction.
8. Explain the advantages of the unit method of instruction. Discuss problems or limitations of this approach.
9. Organize one or more units from a course of instruction in your own occupational field.

REFERENCES

1. Foster, Charles R., *Psychology for Life Today*. Chicago: American Technical Society, 1961.

Chapter 10 INDIVIDUALIZATION OF INSTRUCTION: OPERATION, JOB, AND JOB PLAN SHEETS

Learning is individual. Each person must learn by himself or herself and assume responsibility for his or her own achievements in the learning process. Often related instruction must be individual or individualized.

The learning style of one individual will differ from that of another. Likewise, the rate of learning will differ between and among individuals as a result of native ability, environmental experiences, learning conditions, and motivation of the learner. The amount of skill and knowledge retained over time will also vary with the individual.

Since learning ability, rate, and retention reflect individual differences, the tools for learning and teaching likewise need to be designed for individual differences.

Individual instruction is usually interpreted to mean having one student working with one instructor in a particular course or subject matter area. The individualization of instruction describes the conditions present by which the teacher provides a learning environment for the student or students similar to that present in individual instruction. It is possible to individualize instruction for small groups as well as for single individuals. It is also possible to individualize the instruction for larger groups to better serve individual differences.

Employers hire individuals. Individuals complete tasks and jobs as part of their payroll responsibility. Individuals often work with others to produce a team effort. Even within the team, the contribution of each individual is vital to the success of the total project. Performance is the result of individual learning!

Individualized instruction stimulates independent learning. Independent learning has a tendency to develop a person who is more resourceful and likely to apply self-evaluation to his or her own activities. Assessment by self or by others is an essential step in the mastery of skill, acquisition of knowledge, and the production of products or the rendering of services to an acceptable standard.

Instructional programming, today more than ever before, is oriented toward the individualization of instruction. Solutions to problems of the disadvantaged students as well as the continued motivation of the highly gifted students are directly related to individualizing instruction.

Individualizing of instruction requires an administrative structure that is flexible and designed to permit movement through the instructional program of the educational system at various rates. To achieve individualized instruction a system of instruction is necessary which has clearly defined performance goals, adequate curriculum materials, and performance evaluation designed to effectively assess the individual student's achievements. Achievement must be measured in terms of the performance demands of the payroll jobs of prospective employers.

Individualized Instruction Sheets

Self-paced, self-directed student learning requires appropriate instructional aids. The instruction sheet is a part of the learning package or kit of tools for an effective, but inexpensive system of self-paced instruction. These sheets aid the student and improve the effectiveness of the teacher. One of the results is that teacher-directed learning is replaced by teacher-guided learning. Self-directed student learning has a tendency to result in learner-directed accomplishment. Learner-directed accomplishment tends toward an improved individual self-concept and higher motivation.

Instruction sheets are not designed to replace the teacher. They are aids to teaching in a way similar to the manner that audiovisual aids help the student in the achievement of the identified objectives. Carefully planned and well-written instruction sheets help each individual to make maximum progress at his own rate. Effective instruction sheets combine needed facts and information with specific directions for use by the student. The problems of communications are reduced by the use of simple, short sentences or phrases, concise, and to the point. Sketches, drawings, charts, and illustrations are useful in communicating the concepts to the student.

It is desirable to orient the student to the purpose and the use of instruction sheets from the very beginning. He will then find the transition to manufacturers' literature, factory manuals, and handbooks used on the job to be easy and simple. Instruction sheets supplement demonstrations and group or individual discussions. In addition they provide source materials for ready reference. Instructional materials of this type provide a guide for the doing and the thinking necessary for successful completion of the operation and/or job. Skill in reading and following instructions usually results from this approach to self-directed learning and teaching.

RELATIONSHIP TO CONTENT ANALYSIS

The content analysis of a unit of instruction employs as the vehicles of instruction job sheets, operation sheets, and information sheets.

Students in vocational and technical education classes find learning simplified if the flow of involvement is from "things" to "concepts" rather than from "concepts or theory" to "things." This is one of the differences between the applied psychology of vocational education and that of general education. In vocational and technical education the contact of the student with "things" often actually precedes the concept with the "theory." This approach results in greater relevancy to the student and improved desire to learn.

The structure for teaching a unit consists of:

1. Identifying and listing essential operations.
2. Determining typical jobs for learning and teaching the skills, knowledges, habits, and attitudes identified with the objectives of the unit.
3. Delineating topics of information which need to be presented, clarified, or reinforced beyond the treatment readily available in the textbooks and reference books designed for student use.

The content analysis provides the structure for the development of the self-paced, self-directed components of the learning system. This is usually followed by the construction of the system and consists of the development of operation sheets, job sheets, and information sheets. (Information sheets will be discussed more fully in Chapter 11.)

Instruction sheets are designed for student use and must be written at the appropriate level for reading and comprehension.

Jobs, Operations, and Steps

The selection of typical jobs (sometimes called type jobs) for learning and teaching the skills and knowledges of the unit should precede the development of job sheets or operation sheets. After appropriate jobs have been selected and arranged in order from simple to more complex, then the operations essential to performance of the jobs can be determined. These operations are then arranged in logical order for learning from simple to complex. The order of operations is that order which must be followed to prepare the student to successfully complete the typical job selected.

It is important at this point to remember that:

1. A step is the smallest increment of an operation which advances the work. Two or more steps must be present in an operation.
2. An operation is a unit of work. It may be either mental or manual. Frequently it involves making, servicing, or repairing. It may be compared to the operations in the solution of a problem in mathematics. Operations are usually expressed using words ending in "ing" such as shaping, forming, assembling. Operations are basic to completion of jobs. These are repeated over and over again in the performance of the job. For this reason operations are basic in learning and teaching. An operation consists of two or more steps. Two or more operations make up a job.
3. A job is used to refer to the completed project or it may be used to designate a payroll classification. A typical job for instructional purposes is a completed project for which a person could be paid or that a concern can sell to a customer. A job may be a portion of a completed project reduced in size to produce a manageable teaching increment. The job will consist of two or more operations.

Operation Sheets

An operation sheet provides the sequence to follow in performing a single operation. This operation usually involves both mental and manipulative processes. The operation sheet should be detailed enough so that a student can follow the steps as given and successfully complete the operation. The operation is usually demonstrated by the instructor prior to the assignment of an operation to the student for performance. The operation sheet serves as reference material for the slow student after viewing the demonstration. The fast student often can perform the operation without a demonstration by the instructor using the operation sheet as a guide.

The purpose of an operation sheet is to provide the steps which must be completed to do the operation. **The operation sheet tells how to do the operation.**

(Text continued on p. 163)

EXHIBIT 10–1

Suggested Format for Operation Sheet

School ____________ Operation No.______
Use same No. as in course of study.

Insert name of operation

Sketch or Drawing:	Provide drawing or make sketch if necessary.
Materials:	List materials needed to perform operation.
Tools and Equipment:	List tools and equipment needed to perform job.
Procedure:	State each step in the operation. Steps should be stated clearly and concisely in occupational terms. List steps in proper sequence. Number steps consecutively. Safety and key points should be listed with steps where they apply. Double space between each step.
Checkpoint:	A checkpoint may occur at any step in the procedure where the instructor desires to check the trainee before allowing him to proceed. There may be more than one checkpoint in the operation. Place the word CHECKPOINT in the left margin at the point in the procedure where the trainee is to be checked by the instructor before the trainee is to proceed with the next step.

Some representative occupational areas which might be the subject of operation sheets are:

Occupation	*Operation*
Auto mechanics	Make a compression test on a motor.
Electrical	Thread a piece of conduit.
Machine shop	Sharpen a drill.
Plumbing	Cut pipe with a pipe cutter.
Printing	Ink a press.
Sheetmetal	Wire an edge.
Cosmetology	File fingernails.
Dressmaking	Make a mitered corner in a hem.

Source: The Preparation of Occupational Instructors. Washington, D.C.: Superintendent of Documents, 1966, pp. 105–106.

EXHIBIT 10–2

SAMPLE OPERATION SHEET

How to: Chip with a Cold Chisel | Operation Sheet 18

Course: Beginning Machine Shop Practice | Page 1 of 1 Pages

Introduction:

Chipping is the process of removing metal by means of a cold chisel and hammer. The use of the shaper, milling machine, and planer are more efficient methods of removing metal accurately and rapidly, but the use of a chisel is necessary on many jobs where accuracy is not important and only a small amount of metal is to be removed.

Tools and Equipment:

Cold chisel | Machinist's hammer
Machinist's vise | Goggles

Procedure:

1. Mount the work firmly in the vise; use soft copper jaws if the work has finished surfaces.
2. Hold the chisel with the thumb and fingers of the left hand, so that the head end extends above the hand.
3. Place the cutting edge of the chisel on the surface of the job where the cut is to be made. The chisel should be held at a cutting angle of approximately 45 degrees.
4. Grasp the hammer near the end of the handle so that it can be swung with an easy forearm movement.
5. Strike the head of the chisel with a firm sharp blow.
6. Reset the cutting edge of the chisel on the work and repeat the above steps.

Safety Precautions:

1. Goggles should be worn during the chipping process.
2. Make sure that the chisel head is not "mushroomed," as particles may break off and cause a personal injury.

EXHIBIT 10–3

Sample Operation Sheet

Operation: Observation and Recording

Title of Unit: Practical Nursing Procedures

Field: Practical Nursing

HOW TO COUNT RESPIRATION

Tools and Materials

Watch

Steps	Key Points
1. Count respiration by watching the rise and fall of chest of patient breathing without his being aware of what you are doing.	1. Count immediately after counting pulse while fingers are still on wrist. Do not relax fingers until finished.
2. Count each inhalation and exhalation as one respiration.	2. Count for one minute.
3. Record immediately.	3. If rate per minute is less than 14 or more than 28, report to head nurse.

Safety: None

Questions: 1. Why is it important to count the respirations without the patient being aware of it?
2. How do you determine an abnormal respiration?

EXHIBIT 10–4

Sample Operation Sheet

Operation: Replacing Gas Bottles

Title of Unit: Auto Body Welding

Field: Auto Body and Fender Repair

Tools and Equipment:

- One full bottle of oxygen
- One full bottle of acetylene
- Bottle holding cart
- Regulator wrench

Operation Steps	*Safety*	*Key Points*
1. Turn bottle valve off. Remove regulator from empty oxygen bottle.	Don't bump or drop regulators; they are delicate instruments.	Turn oxygen regulator flange nuts counterclockwise to remove.
2. Turn bottle valve off. Remove regulator from empty acetylene bottle.		Turn acetylene regulator flange nuts clockwise to remove.
3. Replace valve hood on both bottles.	Never move a gas bottle until valve hoods are in place.	
4. Unlatch hold-down chain, remove empty bottles from cart, and place empty bottles in storage.	Don't drop bottles—may damage bottles or injure person handling bottles.	
5. Place full bottles on cart and secure hold-down chain.		
6. Remove valve hoods and store in safe place.	Never move a full gas bottle unless valve hoods are in place.	
7. Replace oxygen and acetylene regulators on bottles.	Never put petroleum lubricant on any part of gas welding equipment.	Install regulators in reverse of operations one and two.

Questions:

1. Why do welding regulators have opposite threads on flange nuts?
2. What may happen if bottle valve is broken off a full bottle of gas?
3. Why do we never put petroleum on any part of gas welding equipment?

A good operation sheet:

1. Is written in clear and concise language.
2. Is arranged in good learning order.
3. Explains and illustrates new terms and nomenclature used.
4. Illustrates steps that may cause difficulty.
5. Identifies key points (elements that will make or break the job).
6. Emphasizes safety and lists applicable precautions.

The following sections are frequently used on the operation sheet:

1. Boxhead with name of school, operation, and unit (identification numbers are useful)
2. Performance goals
3. Introduction to operation
4. Sketch or drawing
5. Tools and materials
6. Procedures and key points (in two columns or combined as one)
7. Checkpoints
8. Questions
9. References

Several formats may be used in writing operations. The one selected will vary somewhat with the nature of the operation and the preference of the instructor. Several acceptable formats are provided as illustrations in Exhibits 10–1, 10–2, 10–3, and 10–4.

Job Sheets

The job sheet provides essential information and the proper sequence to complete the job. In an instructional program it is used in conjunction with the operation sheet. Whereas the operation sheet contains information on how to do the operation, the job sheet may refer to operation sheets previously used and integrates from several operation sheets the essential sequence necessary to perform the job.

The job sheet in industry or in school rarely needs to give directions for doing the job. This the employee should know. It is helpful however for a student in school to have more information relative to the procedure to follow than would normally be provided in industry. But do not duplicate the information already contained in the operation sheet—reinforce it.

The job sheet supplements the instructor's demonstrations and previously completed operation sheets. It serves as a reminder for the student as to the purpose of the job, the materials and equipment required, and the procedure to be followed. (Text continued on p. 169)

EXHIBIT 10–5

THE JOB SHEET

In vocational education, learners are usually assigned actual jobs to perform under varying degrees of supervision. To help ensure success and high standards, the *instructor prepares a job sheet* for each such job. Job sheets are particularly useful in classes where different levels of instruction are being given at the same time.

Some typical jobs for which a learner would need a job sheet are:

Occupation	*Job*
Auto mechanics	Reline and adjust the brakes on a car
Electrical	Install an extra convenience outlet
Commercial food preparation	Make an order of cupcakes
Plumbing	Run in a roof vent
Printing	Print a wedding invitation
Machine shop	Make a flanged bushing
Sheetmetal	Make a section of sheetmetal cornices
Cosmetology	Make a pin curl

As used in the occupation itself, a job sheet may be quite simple, containing a blueprint or sketch and providing only the minimum information needed for the job. In a sense, the tickets, shop blueprint, work orders, and similar items used in an occupation are job sheets.
However, the learner in his early training needs more information than the experienced worker. For that reason the job sheets used in training are designed to help the trainee learn *how* to do the job, as well as to serve as a job assignment. Job sheets used for instructional purposes usually contain:

1. A statement of the job to be done.
2. A list of materials and equipment needed.
3. A procedure outline.
4. Directions for checking the finished work.
5. Pictures, diagrams, working drawings, and sketches to show what is wanted.
6. Pictures, diagrams, and sketches to clarify any anticipated difficulties the learner may have.

Gradually, the job sheets given a learner should become more like the job ticket, work order, or blueprints he will use in the occupation.

Some job form or job sheet should be prepared for each job that will be taught in a course.

Source: The Preparation of Occupational Instructors. Washington, D.C.: Superintendent of Documents, 1966, p. 101.

EXHIBIT 10–6

JOB SHEET FORMAT

School____________

Job No.________
Use same No. as in course of study.

Insert name of job

Sketch or Drawing: Provide drawing or make sketch if instructor deems necessary.

Materials: List materials needed to perform job.

Tools and Equipment: List tools and equipment needed to perform job.

Procedure: State each operation or step in the job. Operations should be stated clearly and concisely.

List operations in proper sequence.

Number operations consecutively.

Safety and key points should be listed with the operations where they apply.

Double space between each operation.

Checkpoint: A checkpoint may occur at any operation in the procedure where the instructor desires to check the trainee before allowing him to proceed.

There may be more than one checkpoint in the job.

Place the word CHECKPOINT in the left margin at the point in the procedure where the trainee is to be checked by the instructor before the trainee is to proceed with the next operation.

Source: The Preparation of Occupational Instructors. Washington, D.C.: Superintendent of Documents, 1966, p. 102.

EXHIBIT 10–7

SAMPLE JOB SHEET

School____________ Job Number 5

Turn Standard Lathe Mandrel

Performance Goals:

1. To turn mandrel to specified size.
2. To turn mandrel with taper per inch specified.
3. To turn mandrel to finish required by print.

General Information and Instructions:

1. Turn mandrel .015″ over print size to allow stock for grinding.
2. Turn mandrel with .0005″ per inch of taper on body of mandrel.

Equipment, Tools, and Materials:

1. Stock
2. Turning, facing tools
3. Counterboring tool
4. Lathe
5. Lathe dog
6. Micrometer
7. Mill file
8. Center drill and drill chuck
9. Rule
10. Steel letters and hammer

Precautions:

1. When center drilling work in a lathe held in a chuck do not extend the work out from the chuck more than 1½ times the diameter of the work.
2. Never leave the chuck wrench in the chuck when setting up work in a chuck.

Procedure:

1. Cut stock adequate to face ends to print dimension.
2. Put four-jaw chuck on lathe.
   Note: Place board on ways before changing from face plate to chuck.
3. Clamp work in chuck, center, and center drill; face end of shaft.
4. Grind tool for counterboring; counterbore to plan.
5. Recheck center hole for ⅔ depth of countersink.
6. Turn shaft end for end; chuck and center; face to length; center drill and counterbore.
7. Remove chuck and install centers and face plate.
8. Set up lathe for straight turning and turn both ends of mandrel to print size.
9. Turn body of mandrel to .015″ over print size with proper taper.
10. Form radii with file to approximate size.
11. Layout flats on both ends of mandrels with point of facing tool.
12. File flats.
13. Stamp size of mandrel on the round portion directly opposite one of the flats on the large end. Stamp size in decimals.
14. Inspect manufactured part for sizes, finish and workmanship. Hand in part.

References:

1. Print No. 3
2. Information Sheet No. 3

EXHIBIT 10–8

Sample Job Sheet

Subject: Building a Halfwave Rectifier Circuit — Job #1

Objective: To build a halfwave rectifier with material listed. Time: approximately one hour.

Schematic:

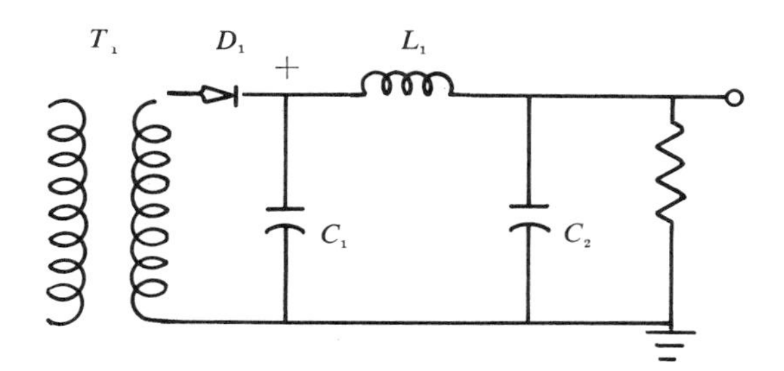

Materials:

1. Circuit board (1)
2. Diode (1) IN54D
3. Transformer (1) T543
4. Coil (1) 10 henry
5. Capacitors (2) 20 μF
6. Resistor (1) 20 K

Tools and Equipment:

1. Long nose pliers
2. Heat sink tool
3. Solder
4. Wire
5. Diagonal cutters
6. Soldering iron
7. Wire brush

Introduction: The halfwave rectifier is the simplest of the power supplies. Its construction and operation should be a big first step forward for you in your study of rectifier circuits.

Procedure:

	1. Mount capacitors	
Checkpoint	2. Observe polarity	(If necessary, see Information Sheets Nos. 1, 2, 3, and 4.)
	3. Mount transformer	
	4. Mount coil	
	5. Mount resistor	
	6. Mount fuse	
	7. Mount switch	
	8. Mount diode	
Checkpoint	9. Use heat sink tool	
Checkpoint	10. Observe polarity	

References:

1. S.S.G. Block IV Power Supplies and Amplifiers
2. A.F.M. 101–8

Study Questions:

1. Why did you use heat sink tool?
2. Why is diode polarity important?
3. Would capacitors work either way?
4. Where did you place the iron?

EXHIBIT 10–9

SAMPLE JOB SHEET

Data Processing Area

Operation 2—Operation of the 1051 Card Reader

Objectives:
Each student will be able to properly operate the 1051 Card Reader.

General Information and Instructions:
1. Turn on power to the Card Reader.
2. Perform a Lamp test at the diagnostic panel.
3. Load cards into the input hopper properly.

Equipment, Tools, and Materials:
Console Operator's General Reference Manual
1051 Card Reader
Program Card Decks

Precautions:
1. Insure that all panels on the Card Reader are securely fastened in place.
2. Place the card weight on top of the program deck in the hopper, else the machine may jam before all cards are read into the processor.
3. Make certain that maintenance work is not being performed on the unit.

Procedures:
If the 1050–II System power is on and the POWER ON switch is not lit on the reader control panel. (POWER OFF diagnostic indicator lit.)
1. Depress the POWER ON switch. The POWER ON indicator and the READY indicator will light.
2. DEPRESS and hold READY switch to accomplish a lamp test on all control panel lamps. If any lamps are defective, inform the field engineer.
3. Depress the OFF-LINE switch (lights) to divorce the card reader from the Central Processor.
4. Visually check the read ready station for a stray card. If a card is present in the read ready station, depress MOM portion of MANUAL FEED pushbutton to clear the card feed track of any cards.
5. Load cards in input hopper 9-edge leading face down.
6. Depress OFF-LINE (extinguishes) to place the unit on-line under control of the Central Processor.
7. Depress READY to extinguish any false diagnostic indications, i.e., MISFEED, etc.

References
Use Figures 4–2, 4–3, and 4–4 as listed in the SKETCHES section of Operation Sheet 2.

The job sheet usually contains:

1. The name of the job
2. Drawings or references to prints
3. Materials
4. Tools and equipment needed
5. General instructions
6. Order of operations
7. Checkpoints

The purpose should be identified in terms of measurable performance goals and listed on the job sheet. Questions and references are helpful as additional learning devices.

The detailed procedures in doing each operation were previously provided in the operation sheet; therefore, the job sheet need only provide the sequence in correct order. References to operations are frequently made for the convenience of the student who needs to refresh his memory. To illustrate the content and format of job sheets the samples in Exhibits 10–5, 10–6, 10–7, 10–8, and 10–9 have been selected.

Job Plan Sheets (Student Work Plans)

As students enter the more advanced phases of instruction, it is highly desirable that opportunities for them to plan their own work be provided. The student work plan provides such opportunities. This is developed by the student under the guidance of the instructor. The student work plan is in fact a job sheet prepared by the student himself. This sheet needs to supply the same kinds of information as a regular job sheet.

Three important reasons exist for having students complete the student work plan. These are:

1. To prepare a plan of action that will result in the development of skill.
2. To provide practice in analyzing and planning jobs.
3. To enable the instructor to follow the student's progress in doing the job.

Other reasons for the student work plan are:

1. To be sure the student knows the proper sequence of operations before beginning to work on the job.
2. To be sure the student has solved the practical problems required for the completion of the job.
3. To save the time of the instructor and the student while in the shop or laboratory.

EXHIBIT 10–10

STUDENT WORK PLAN (JOB PLAN SHEET)

School____________ Job Plan No.________

Trainee's Name____________ Use same No. as on progress chart.

Insert name of job

Sketch or Drawing: If instructor deems necessary, trainee makes sketch or drawing as directed.

Materials: List materials needed to perform job.

Tools and Equipment: List tools and equipment needed to perform job.

Procedure: State each operation or step in the job.
Operations should be stated clearly and concisely.
List operations in proper sequence.
Number operations consecutively.
Safety and key points should be listed with operations where they apply.

Checkpoint: A checkpoint may occur at any operation in the procedure where the instructor desires to check the trainee before allowing him to proceed.

There may be more than one checkpoint in the job.

Instructor draws a red line in the left margin at the point in the procedure where the trainee is to be checked by the instructor before the student is to proceed with the next operation.

Estimated time________hours.
(Determined by trainee, subject to instructor's approval)

Actual time________hours.
(Actual hours to complete job)

____
Approved by instructor
(Initialed or signed before work is started)

*Source: *The Preparation of Occupational Instructors.* Washington, D.C.: Superintendent of Documents, 1966, p. 104.

EXHIBIT 10–11

SAMPLE JOB PLAN SHEET

*Trainee's Name*______________________________ Job Plan No._______

Name of Job ____

Sketch or Drawing:

Materials:

Tools and
Equipment:

Procedure:

*Estimated Time*_______

Actual Time _______

____
Approved by Instructor

4. To provide a record of the work experience of the student.
5. To provide a basis for future assignments.

The format and headings for the student work plan (job plan sheet) may vary, but the basic content and purpose are the same as shown in Exhibit 10–10. A sample job plan sheet is shown in Exhibit 10–11.

Summary

All learning is individual. Each person must assume responsibility for his (or her) learning.

Instructional programming in schools is moving rapidly toward individualizing of the instructional process. Self-paced instructional programs necessitate either a "software" or a "hardware" system to aid the learner. A simple inexpensive plan can be provided through the use of instruction sheets. Instruction sheets are appropriate for individual, small group, or larger group instruction.

An operation contains two or more steps. A job contains two or more operations. An operation sheet lists the essential steps together with other information needed to perform the operation.

The job sheet gives the sequence to follow in completing the job. Other helpful information is included on the job sheet.

The student work plan is a job sheet prepared by the student for performance of the job.

QUESTIONS AND ACTIVITIES

1. Discuss the advantages of individualized instruction. Limitations.
2. What features are needed in a self-paced instructional system?
3. What are the advantages of using instruction sheets?
4. Explain:
   A. Step
   B. Key point
   C. Operation
   D. Job
5. What kind of information should be provided in the:
   A. Operation sheet?
   B. Job sheet?
   C. Student work plan?

6. Describe the format of the:
    A. Operation sheet
    B. Job sheet
    C. Student work plan
7. Develop for a unit of your occupation one or more:
    A. Operation sheets
    B. Job sheets
8. Why is a student work plan a good learning and teaching tool?
9. At what point in the instructional program should students become accustomed to the development of their own student work plans rather than depend upon job sheets?
10. Can operations be developed for levels of occupations heavily dependent on mental rather than skill processes? Discuss.

Chapter 11 INDIVIDUALIZATION OF INSTRUCTION: INFORMATION SHEETS

Information sheets grow out of the needs identified by analysis. Related or applied information often must be provided to assist the learner in relation to operations to be performed. In such cases, an explanation is given by the instructor at the time of the demonstration of an operation. Available textbooks and reference books provide additional information for the student. However, many times needed information is not readily accessible to the student in available textbooks or reference books; therefore, information sheets are important sources of related information.

Conditions Necessitating Development of Information Sheets

On occasion, only one copy of a pamphlet or brochure is available. In order to expedite transmitting the information to the students the vital content may be abstracted and provided in the form of an information sheet.

Sometimes the treatment of the topic is provided by several authors in a limited manner. None of the books may provide the detailed description or information sought by the teacher for the students. The teacher may rewrite into one information sheet the ideas expressed in part by several authors.

In other cases the treatment of the topic by available reference books is too advanced for the comprehension of the students. It is then necessary to translate the meaning into words which the students can understand. This is logically done through the process of writing an information sheet and adding illustrations, pictures, charts, or diagrams.

In some cases information is needed from the literature of manufacturers or vendors in which only part of the information is desired. In other cases, information sheets may be used if the number of copies of the literature is limited.

The content for information sheets may be extracted from periodicals, pamphlets, and booklets to provide a basis for better understanding for the student learning.

KINDS OF INFORMATION PROVIDED

Three different kinds of information sheets are usually provided to students. These have often been described as:

1. *Must-know* information sheets
2. *Nice-to-know* information sheets
3. *Guidance* information sheets

The *must-know* information sheets may be thought of as those bits of related or applied information which the student must have before he can effectively proceed with the operation or job assigned.

The *nice-to-know* information sheets provide enrichment to the learning experience. These bits of information are not absolutely essential but are useful for better understanding.

Guidance information lends motivation and realistic understanding to the student concerning the world of work and the opportunities of the person in the world of work who has achieved the competencies essential for entry. Such information sheets may deal with such topics as: job opportunities, job descriptions, job requirements, working environment, interpersonal relationships, opportunities for advancement, etc. (as shown in Figure 11–1).

RELATIONSHIPS TO OPERATION SHEETS AND JOB SHEETS

Information sheets are frequently used in conjunction with operation sheets. The content analysis chart provides a source for identification of information sheets needed with operation sheets designed to assist the learner in doing the operation.

Sometimes information sheets are coupled with operation sheets. How-

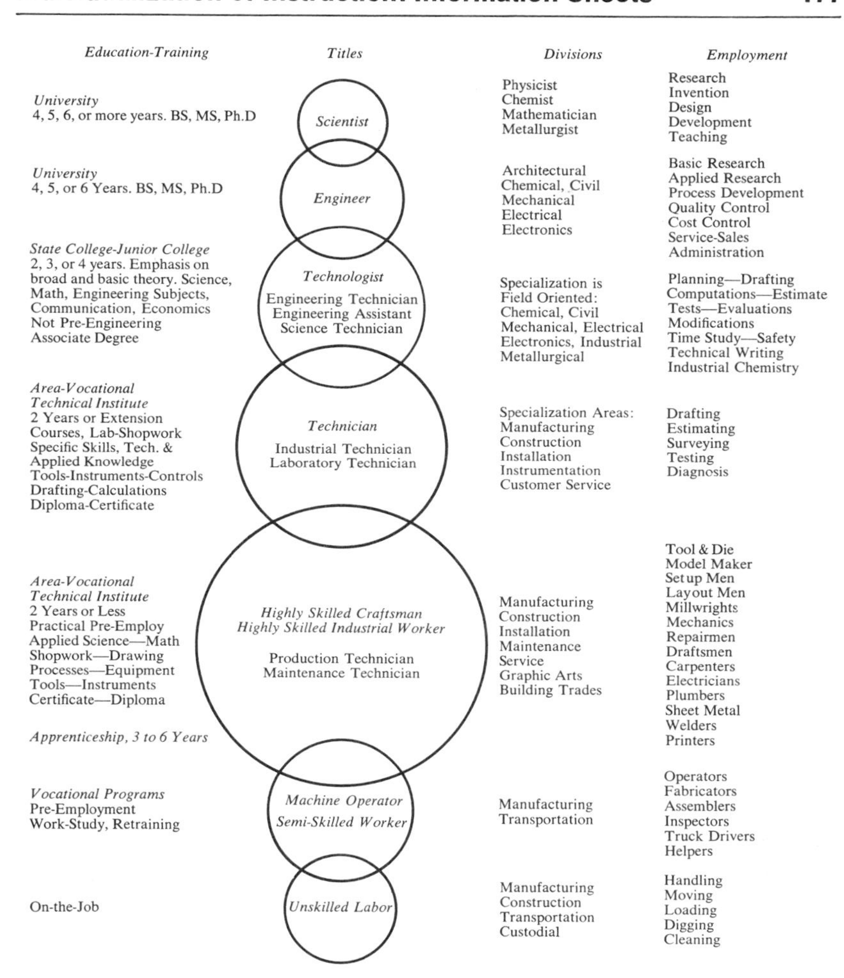

Source: John Butler, Dunwoody Institute, Minneapolis, Minnesota

FIGURE 11–1

GUIDANCE INFORMATION SHEET
PROFESSIONAL, TECHNICAL, CRAFT, AND INDUSTRIAL OCCUPATIONS

ever, it is important that the "must-know" kind of information be provided at the time the operation is assigned. "Nice-to-know" information sheets and guidance information sheets may be provided to strengthen the learning process in connection with any of the shop, laboratory, or related class work in vocational and technical education.

Topics for Information Sheets

Since information sheets are intended to supplement readily available sources such as textbooks and reference books they need not be written on all topics included in the course of study. In fact, the time of the instructor is too valuable to be used for writing information sheets on topics already adequately covered. After reviewing coverage in the sources available to the students, the instructor makes the determination of the topics to be developed into information sheets.

Topics such as the following may need to be treated by separate information sheets:

Occupation	*Information*
Auto mechanics	The principle of the differential in an automobile
Electrical	The principle of the transformer
Plumbing	Cast iron pipe—types, sizes, and uses
Printing	The point system of measurement
Commercial food preparation	The action of yeast in bread dough
Cosmetology	The purpose and use of astringents
Dressmaking	Linen—what it is and where to use it

(1, p. 99)

Sample information sheets are given in Exhibits 11–1, 11–2, 11–3, and 11–4.

EXHIBIT 11–1

INFORMATION SHEET

School ____________________ Information Sheet No. ______
Use same No. as in course of study.

Insert title of information to be presented

Information: Written to suit level of trainee.
Sentences and paragraphs should be concise.
Illustrations should be used where they will assist in clarifying the information.
Material should be organized and presented in a logical sequence.
Material should be of sufficient length and complexity to challenge the trainee.

Source: The Preparation of Occupational Instructors. Washington, D.C.: Superintendent of Documents, 1966.

EXHIBIT 11-2

SAMPLE INFORMATION SHEET

Subject: Shadow Mask Assembly

Title of Unit: Pre-set Adjustments

Field: Color Television Repairman

1. This mask has 330,000 etched holes. There are one million phosphor dots on the inside of the kine face plate ahead of it. The shadow mask is inside of the kine about 9/16 of an inch behind the face plate. There is one hole in the shadow mask for each dot TRIAD of red, green, and blue dots.
2. Each of the three beams from the electron guns must pass through and converge at the holes in the shadow mask and strike the appropriate phosphor dot.
3. Not all of the beam current can pass through the holes in the shadow mask and strike the dots. The electron beams are larger in diameter than the shadow mask holes. Part of the electrons in the beams are therefore collected by the shadow mask. The shadow mask is made of the conductive material and is connected to the second anode. This part of the beam current is returned to the HV power supply.
4. The three electron beams must CROSS AND CONVERGE and PASS through the holes at the shadow mask. In doing this, if the receiver is properly adjusted, the appropriate phosphor dots will be struck and will produce the very small red, green, and blue lights. These lighted dots may be observed with a microscope on the face of the TRI-COLOR kinescope.

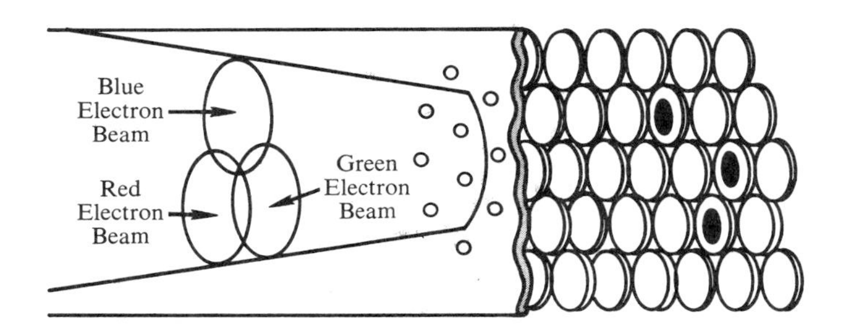

EXHIBIT 11–3

SAMPLE INFORMATION SHEET

Information Sheet No. 1–1 _______

ELECTRONIC FORMULAS

Introduction: Listed below are a few of the more common formulas you will be utilizing during your training in Electronic Fundamentals. Other information sheets will be issued, covering additional formulas, as you progress through the course. Save them and at the completion of your training program, you will have a complete and ready reference list.

Resistance Formulas

In Series $R_t = R_1 + R_2 + R_3 \ldots$ etc.

In Parallel $R_t = \dfrac{1}{\dfrac{1}{R_1} + \dfrac{1}{R_2} + \dfrac{1}{R_3} \ldots \text{etc.}}$

Capacitance

In Parallel $C_t = C_1 + C_2 + C_3 \ldots$ etc.

In Series $C_t = \dfrac{1}{\dfrac{1}{C_1} + \dfrac{1}{C_2} + \dfrac{1}{C_3} \ldots \text{etc.}}$

Ohm's Law for DC Circuits

$I = \dfrac{E}{R}$ $R = \dfrac{E}{I}$ $E = IR$ $P = EI$

Where: I = current in amperes E = potential across R in volts
R = resistance in ohms P = power in watts

Peak, rms and Average AC Values of E & I

Given	To Get		
Value	Peak	rms	Avg.
Peak		0.707 × Peak	0.637 × Peak
rms	1.41 × rms		0.9 × rms
Avg.	1.57 × Avg.	1.11 × Avg.	

EXHIBIT 11–4

INFORMATION SHEET

Using wires to check the thread size

Objective: To enable the student to check his threads using the wire method.

$$M = D + 3G - \frac{1.5155}{N}$$

$$G = \frac{.57735}{N}$$

M = measurement over the wires
D = major diameter of the screw
N = number of threads per inch

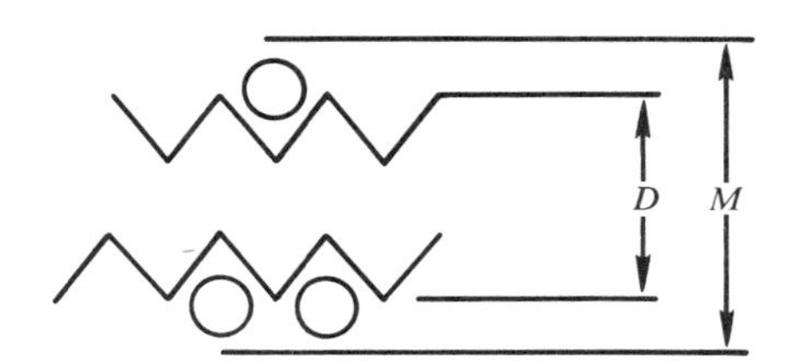

Two wires are placed on one side and one wire on the other side. When the threads are right, the measurement over them is "M"

Summary

An information sheet contains facts desirable for addition to the operation sheets and other text and reference materials available to the student. The most common information sheets contain knowledge essential to progress or performance of an operation or job similar to that found in a payroll job. Other information sheets are designed to provide enrichment of the learning experience or guidance desirable for student decision making and career planning.

The information sheet should be synchronized with the other instructional sheets with which it relates. Since the information sheets are to be studied rather than to direct the activities of the student, the format used is less formal. The information sheet should be easy to read, clear, and concise. It should also be interesting to read and should hold the attention of the reader. Motivation and stimulation can be achieved through the use of pictures, illustrations, charts, sketches, diagrams, and tables.

The information sheet may have identified objectives, questions, and references.

QUESTIONS AND ACTIVITIES

1. Explain the relationship of information sheets and the content analysis chart.
2. Are information sheets needed for each operation sheet? Explain.
3. What conditions suggest the need for writing information sheets?
4. Explain the functions of:
   A. *Must-know* information sheets
   B. *Nice-to-know* information sheets
   C. *Guidance* information sheets
5. Suggest several topics for information sheets in your own occupational field.
6. Describe several common types of information sheets.
7. Select from a content analysis one or more topics for essential information sheets. Write the information sheet(s). Illustrate as needed.
8. Discuss types of formats for information sheets.

REFERENCES

1. *The Preparation of Occupational Instructors.* Washington: D.C.: Superintendent of Documents, 1966.

Chapter 12 ASSIGNMENT SHEETS

A clear indication of what the student is to do is vital to the learning process. One of the most effective ways of achieving this function is through a well-planned assignment sheet. Assignment sheets are important aids in teaching related courses.

Assignment sheets can be developed for the entire course, a unit of the course, or a particular lesson of the unit.

The assignment sheet is one of the most valuable of the instructional sheets both from the viewpoint of the student and the instructor. The assignment sheet helps the student to plan ahead and use his or her time most effectively. It also forces the instructor to do more long-range planning.

Relationship to Operation Sheets, Job Sheets, and Information Sheets

Assignment sheets are developed using information gained through analysis concerning the "doing" and "knowing" activities of the students. These sheets need to cover the classes of related and applied subjects as well as laboratory and shop classes.

The assignment sheet can be used equally effectively with individual and group instruction. This sheet must provide the learner with definite instructions as to what he is to do and how he is to do it. Often a number of problems or questions are given to aid the student and the teacher in assessing the degree of achievement. In essence, it supplements the instructions given orally by the teacher and provides a ready reference guide for the student relative to the expectations of the lesson, unit, or course. Typical examples along with suggested format are given in Exhibits 12–1, 12–2, and 12–3.

Course or Unit Assignment Sheet

The purpose of the course assignment sheet is to give the student a firm concept of the requirements of the course. This may be supplemented with additional assignments later if necessary. It could also be modified if the progress of the student or the class as a whole is such that adjustment is desirable.

The course or unit assignment sheet provides such basic information as:

1. Course description or unit description
2. Objectives
3. Reading assignment
4. Textbook or textbooks
5. References
6. Course activities
7. Information relative to other requirements

The course or unit assignment sheet has some of the same characteristics as the acceptance of a job to be completed in a jobbing shop. The performance requirements are identified, the quantity of production is established, and the expected results in terms of evaluation may also be delineated.

It has several advantages:

1. Permits students to know from the very beginning of the course what is expected of them in terms of quantity and also, sometimes, quality of production showing outcomes of learning.
2. Permits students to achieve more effectively the objectives through more effective scheduling of time and utilization of limited resources.
3. Saves the instructor's time and efforts often otherwise required to clarify and repeat oral assignments.

Exhibits 12–4 and 12–5 show sample course assignment sheets for two different curriculums. (text continued on page 191)

EXHIBIT 12–1

The Assignment Sheet

Assignment sheets can be used to:

1. Present new related information by having the learner study it in references.
2. Help learner apply information that he is learning. Such sheets provide the element of repetition so essential in the development of judgment and the ability to use information. In general, assignment sheets are most effective when they are used soon after presentation of information.

Examples of assignments in several representative occupations are:

Auto mechanics: Take specific gravity and voltage readings of several storage batteries and describe the condition of each from the data obtained.

Electrical installation: Solve the assigned problems by means of Ohm's Law.

Commercial art: Collect several samples of printing that illustrate the difference between the optical and the geometrical center of the page.

Foundry work: Calculate the estimated weights of the iron castings which are to be made from the accompanying drawings.

Printing: Figure the cost of the stock for the jobs described.

Dressmaking: Examine the samples of textiles furnished and fill out the data requested in the space provided.

Assignment sheets may involve:

1. New concepts to be learned
2. A series of questions to be answered
3. An experiment to be performed
4. Some problems to work out
5. Bills of material to make up
6. Drawings to analyze
7. Data to study and interpret
8. Procedures to plan
9. An observation or investigation to be made and data to be recorded

An assignment sheet should provide examples or other guidance to help the learner succeed. For example, if there are several problems, at least one of them should be solved with the correct approach clearly shown.
The purpose of an assignment sheet is to get a learner to do something. For that reason it is important for such a sheet to motivate the learner, convincing him that he will benefit by completing the assignment.

Source: The Preparation of Occupational Instructors. Washington, D.C.: Superintendent of Documents, 1966, p. 97.

EXHIBIT 12–2

Assignment Sheet Format

School ______________________________ Assignment Sheet No. _____
Use same No. as in course of study.

Insert title of assignment

Introductory Information:	Short, concise statement to motivate the trainee to complete the assignment.
Assigned Readings:	Listing of all printed matter and other resources to be used by the trainee in completing the assignment. List the facts of publication: author, title, publisher, date, and page numbers covering assignments.
Questions, Problems, or Activity:	The assignment should adequately cover the specific subject in this assignment sheet. Questions and problems should be stated clearly and concisely. The trainee should clearly understand what he is to do.

Source: The Preparation of Occupational Instructors. Washington, D.C.: Superintendent of Documents, 1966, p. 98.

EXHIBIT 12–3

Sample Assignment Sheet

Color Mixing

Introductory: To isolate a color defect to a particular stage in the chrominance section, one must be able to interpret color deficiencies as seen on the TV screen. To increase proficiency in this area, you will do the following problems. Use your chromaticity chart.

Assigned Readings: Kiver, Milton S., *Color Television Fundamentals.* McGraw-Hill Book Company, New York, 1964. Read pp. 1–10.

Problems:

1. White − blue = ____________________
2. White − red = ____________________
3. White − green = ____________________
4. Blue + red = ____________________
5. Green + blue = ____________________
6. Red + green = ____________________
7. Yellow + blue = ____________________
8. Determine the complementary colors of the following:

    a. Red____________________

    b. Blue____________________

    c. Cyan____________________

    d. Yellow____________________

    e. Green____________________

    f. Orange____________________

EXHIBIT 12–4

SAMPLE COURSE ASSIGNMENT SHEET

SCHOOL__________ TRANSISTOR ELECTRONICS I

COURSE INFORMATION

Course Description: This course will introduce the student to the fundamentals of solid state circuits using the bi-polar transistor. Mathematical analysis will be used to illustrate some of the basic circuit considerations. Subject matter begins with the atomic structure of crystals as it applies to transistors, then to the characteristics of the transistor, and finally how it is used in several types of circuits. The laboratory workbook offers an orderly procedure for the student to apply principles studied and discussed in the classroom.

Course Objectives: The student who successfully completes this course will be able to:

1. Explain the basic atomic structure as it applies to solid state crystals used in transistors.
2. Identify the basic characteristic of bi-polar transistors.
3. Solve problems to demonstrate his understanding of the characteristics of the three circuit configurations. This will include biasing, amplification, input considerations, and temperature effects.
4. Complete laboratory experiments that demonstrate his knowledge of principles studied in the classroom.
5. Score an accuracy of 75% or better on his mid-term and final examinations.

Topic	Veatch pages	Tomer pages	Information Sheet No.
1. Semi-conductor materials	1–11	1–18	TRE— 1 & 2
2. Junction Transistor	12–54	19–36	TRE— 3–6
3. Characteristic Curves	35–43		TRE— 7–15
4. Circuit Configurations	44–68	27–30 42–47	TRE—16–19
5. Common Emitter Amplifier	69–92	69–73	
6. Common Base Amplifier	112–120		
7. Common Collector Amplifier	93–111		TRE—20

EXHIBIT 12–4 continued

Textbooks:

Tomer, Robert B., *Semiconductor Handbook.* Indianapolis: Howard W. Sams & Co., Inc. Publishers, 1970. ($2.95)

Veatch, Henry C., *Transistor Circuit Action.* New York: McGraw-Hill Book Company, 1968. ($5.95)

Laboratory Workbook:

DeVry Workbook, *Transistor Circuits.* Chicago: DeVry Industries, Inc., 1968. ($2.85)

References:

Dahlen, Phillip, *Semi-Conductors from A to Z.* Blue Ridge Summit, Pa.: Tab Books, 1970.

Department of the Army, *Basic Theory and Applications of Transistors.* Dover Publications, 1963.

Stoffels, Robert E., "Let's Talk Transistors," *Q S T,* November, 1969, to May, 1970. (Seven-part series published each month in the magazine.)

EXHIBIT 12–5

SAMPLE COURSE ASSIGNMENT SHEET

SCHOOL____________ ELECTRONIC DEVICES

COURSE INFORMATION

Course Description: A study of electronic devices; nomenclature, characteristics, materials, circuits, and how devices work. The subject matter includes vacuum tubes as well as solid state devices. Mathematic tools developed in Mathematics for Electronic Technicians are utilized in a series of experiments strongly laboratory oriented.

Course Objectives: Students who successfully complete this course will be able to:

1. Identify and explain nomenclature of vacuum tubes and solid state devices.
2. Identify and explain characteristics of vacuum tubes and solid state devices.
3. Identify and explain the use of vacuum tube and solid state devices applicable to communications, circuits, and systems with a 90% accuracy.

EXHIBIT 12–5 continued

Topic	Kloeffler, Horrell, and Hargrove, Jr. pages	Ryder pages
1. Introductions. Orientation to course, objectives, and activities		
2. Vacuum Tubes	105–169, 178–189	91–112
3. Photo Tubes	169–174	
4. Gas Tubes	387–403, 404–448	
5. Cathode-Ray Tubes		24–36
6. Semi-Conductor Physics	15–42	37–62
7. Solid State Devices	43–52, 204–265	164–213
8. Special Purpose Devices	52–103	77–89

Textbooks:

Kloeffler, Royce G., Maurice W. Horrell, and Lee E. Hargrove, Jr., *Basic Electronics,* Second Edition. New York: John Wiley & Sons, Inc., 1963.

Ryder, John D., *Electronic Fundamentals and Applications,* Third Edition. Englewood Cliffs, New Jersey: Prentice-Hall, Inc., 1964.

References:

Brophy, James J., *Semiconductor Devices.* New York: McGraw-Hill Book Company, 1964.

A. B. Dumont Laboratories, *The Cathode-Ray Tube and Typical Applications.* Clifton, N.J., 1948.

Dunlap, W. C., Jr., and W. Crawford, *An Introduction to Semiconductors.* New York: John Wiley & Sons, Inc., 1957.

Kittel, Charles, *Solid State Physics.* New York: John Wiley & Sons, Inc., 1956.

Richardson, O. W., *Emission of Electricity from Hot Bodies,* Second Edition. New York: David McKay Co., Inc., 1921.

Schockley, W., *Electrons and Holes in Semiconductors.* New York: Van Nostrand-Reinhold Company, 1950.

Seely, S., *Electron-Tube Circuits.* New York: McGraw-Hill Book Company, 1950.

Spangenberg, K. R., *Vacuum Tubes.* New York: McGraw-Hill Book Company, 1948.

Wilcox, Glade W., *Basic Electronics.* New York: Holt, Rinehart & Winston, Inc., 1960.

EVALUATION OF INSTRUCTION SHEETS

The form given in Exhibit 12–6 is designed to assist students in evaluating the instruction sheets which have been prepared.

EXHIBIT 12–6

CRITERIA FOR GENERAL EVALUATION OF INSTRUCTIONAL SHEETS

Name______________________________Date____________

Instructional Sheet____________________________________
(Type) (Title)

Item	5 Highest	4	3	2	1 Lowest	Comments
I. Content						
A. Appropriate for purpose						
B. Simple, clear, and concise language						
C. Adequately illustrated						
D. Arranged in good learning order						
E. Other:____________ ____________						
II. Format and Style						
A. Suitable heading(s) and title						
B. Good arrangement						
C. Neat and easy to read						
D. Other:____________ ____________						

Total Score____________________

Summary

Assignment sheets are helpful to both the student and teacher. These are methods of conveying to the student what he is to do in the course, in the unit, or in the lesson. After the instructor presents the sheet to the student, opportunities for the student to clarify points needing explanation should be provided.

Assignment sheets aid in:

1. Budgeting time
2. Long-range planning
3. Better use of resources
4. Avoiding confusion as to requirement and expectations

The purpose of the assignment sheet is to get the learner to act. These sheets should be:

1. Clearly written
2. Expressive in directions
3. Specific as to functions to be performed
4. A guide as to what and when

Assignment sheets are one of the most helpful of the instructional sheets. They probably should be the first prepared so the overview of responsibilities is more clearly identified and expressed.

QUESTIONS AND ACTIVITIES

1. Why should instructors prepare assignment sheets?
2. Explain the functions of the:
   A. Course assignment sheet
   B. Unit assignment sheet
   C. Lesson assignment sheet
3. List the characteristics of effective assignment sheets.
4. Establish the criteria for evaluating assignment sheets.
5. How do assignment sheets help the:
   A. Student
   B. Instructor
   C. Supervisor
6. Motivation and development of responsibility are important elements in the learning process. Explain how the assignment sheet serves a vital function relative to each of these factors.
7. How do assignment sheets help to individualize learning and teaching?

8. Select from your occupation and curriculum a specific course. Prepare a course assignment sheet.
9. For one or more of the units of a course develop a unit assignment sheet.
10. Plan a lesson assignment sheet for a lesson of:
    A. A related course
    B. A shop or laboratory course

Part III PREPARING TO TEACH

Chapter 13 BEFORE YOU TEACH—PLAN

The plan or the blueprint precedes action in business and industry. In the teaching profession, also, the most effective learning takes place when appropriate planning and preparation have been done prior to beginning the teaching–learning process in the classroom or shop.

A story has been told that a famous doctor was once asked how he would proceed if he were given an hour to perform an important operation on a patient. He said, "I would take thirty minutes of that hour and plan carefully what to do."

Preparation to teach involves the student and the subject matter that is to be taught to that student. Schools exist only to serve the needs of students—to help them learn more effectively and more rapidly.

Learning includes the acquiring of skills, knowledges, habits, and attitudes through many facets. Like the points of a diamond each may reflect light. These reflections may originate through reading, listening, discussing, solving of problems, simulation activities, and other experiences. Learning requires doing. The doing must be meaningful to the student. It must integrate his interests with the objectives of the educational program.

Teaching includes planning, presenting of opportunities for learning, and measuring accomplishments against the predetermined goals.

While the amount of student accomplishment and the rate of learning will vary from student to student, success of the educational program rests upon how much is learned. Success in preparing the student for entry-level jobs demands achievement of a minimum standard of accomplishment. The success of the teacher can be measured indirectly through the measurement of the accomplishments of the student. While some will deny the truth of the statement, "If the learner hasn't learned, the instructor hasn't taught," it is still fundamentally true.

Sometimes the learning expectations are rather unrealistic. The story of "The Animal School" (source unknown) illustrates this point:

> "The Animal School"
>
> Once upon a time, the animals decided they must do something *heroic* to meet the problems of "a new world." So they organized a school.
>
> They adopted an activity curriculum consisting of running, climbing, swimming, and flying. To make it easier to administer the curriculum, *all* the animals took *all* the subjects.
>
> The duck was excellent in swimming, in fact better than his instructor; but he made only passing grades in flying and was very poor in running. Since he was slow in running, he had to stay after school and also drop swimming in order to practice running. This was kept up until his web feet were badly worn and he was only average in swimming. *But average was acceptable in school, so nobody worried about that except the duck.*
>
> The rabbit started at the top of the class in running, but had a nervous breakdown because of so much makeup work in swimming.
>
> The squirrel was excellent in climbing until he developed frustration in the flying class where his teacher made him start from the ground up instead of from the treetop down. He also developed "charlie horses" from overexertion and then got C in climbing and D in running.
>
> The eagle was a problem child and was disciplined severely. In the climbing class he beat all the others to the top of the tree, but insisted on using his own way to get there.
>
> At the end of the year, an abnormal eel that could swim exceedingly well, and also run, climb, and fly a little had the highest average and was valedictorian.
>
> The prairie dogs stayed out of school and fought the tax levy because the administration would not add digging and burrowing to the curriculum. They apprenticed their child to a badger and later joined the groundhogs and gophers to start a successful private school.
>
> Does this fable have a moral?

How the Student Learns

People learn at different rates with varying degrees of efficiency and comprehension. The impressions are conveyed to the brain where the signals are translated into meaningful concepts. The ability to learn can be improved just as knowledge and skill can be acquired. Awareness of the laws that govern the development of physical and mental habits is helpful in both learning and teaching. Motivation affects not only the rate of learning but also the retention of that which has been learned.

THE SENSES THROUGH WHICH WE LEARN

Sight—We acquire most of our knowledge through the sense of sight. Observation of action and the study of drawings, diagrams, models, and pictures are indispensable in trade training. (Merely reading printed words is considered as being more related to hearing than sight.)

Hearing—Through hearing we are able to learn from the experiences of others. It also enables us to receive instructions and to recognize the proper operation of tools, machines, and the like.

Touch—Through the sense of touch we become aware of the quality and texture of materials, degree of roughness and smoothness, heat and cold, and, to some degree, the shapes of objects.

Smell—The sense of smell is important, to a limited extent, in several trades, chiefly in recognizing materials, chemicals, and the like.

Taste—The sense of taste is perhaps the least used of all the senses, particularly in mechanical trades. Of course, it is important in trades in which foods and drugs are processed.

SOME TRAINING PRINCIPLES

If we are to consider how to train, it would be helpful first to understand something of how men learn.

We are told that we retain approximately:

10 percent of what we read
20 percent of what we hear
30 percent of what we see
50 percent of what we see and hear

The individual in these situations may be relatively passive; he is "receiving." These percentages are clearly only approximations, but they do indicate where emphasis in training should be placed. We shall be much more certain of this, however, if we can find some practical proof. What, for example, have we found out about learning in the vast laboratory of industrial training?

1. Men learn faster by seeing and hearing than by hearing alone.
2. Men learn still faster when doing is added to seeing and hear-

ing. It is doing which makes learning permanent.

3. Men tend to remember more of what they did in training than what they were told in training.
4. Thus, men should be trained for positions under conditions that are as nearly like the actual job as possible. (1, p. 69)

SOME LAWS THAT GOVERN THE DEVELOPMENT OF PHYSICAL AND MENTAL HABITS (SKILLS)

The following are fairly accurate statements of the way in which human beings establish associations and bonds that are finally automatic or semi-automatic habits (skills) in using their bodies and their minds.

1. The mind is a neural switchboard.
2. Through experiences it sets up its own associations of ideas and actions.
3. By practice an association may be developed into a bond.
4. By further repetition a bond may be developed into a habit—a more or less automatic or semi-automatic action of the switchboard.
5. A habit can only be developed by repetition—practice.
6. The effective way to develop any habit of skill in doing or thinking is to practice it correctly from the start.
7. Until it has been fully established a habit should always be practiced in the same way.
8. We carry from one experience in learning to another only the associations, bonds, or habits that we have developed on the first experience.
9. A habit should always be practiced under the conditions under which it is to be used.
10. Habits wane from disuse.
11. Once acquired a habit which has waned is more easily revived than learned in the first instance.
12. The only way to break a bad habit is to practice a correct one in its place.
13. A habit is improved by practicing it against a standard.
14. A habit needs to be practiced until it has become fully established.
15. In the development of a habit (skill) the learner does not steadily progress but moves forward by alternate periods of rest and advancement sometimes called plateaus of learning.
16. A habit is learned most efficiently when it is practiced under the direction of an instructor who is himself *master* of the habit.
17. Just as efficient habits of doing physical work may be developed so may efficient habits of using your mind in thinking.
18. Any habit is developed more quickly and efficiently by a learner who is interested in acquiring it.
19. One way to become resourceful in the use of any habit (skill in

thinking or doing) is to practice it under a wide variety of situations.

20. The only measure of any habit (of skill in thinking or doing) is performance—the ability to use it. (Source unknown)

SOME IMPORTANT FACTORS IN LEARNING IN WHICH INDIVIDUALS DIFFER

The good supervisor realizes that individuals differ in their mental capabilities as in their physical makeup and makes provisions for adapting his training techniques to best fit the learning abilities of each individual worker. Following are some factors which greatly influence a person's ability to learn and which the teacher must recognize in "individualizing" his instruction.

Interest—Without worker interest, instruction is more or less futile. A person learns well those things in which he has a vital, sustained interest. On the other hand, he rebels and ceases to learn when he is bored or fails to see a personal benefit in learning the lesson.

Intelligence—Briefly stated, intelligence is the ability to respond quickly and successfully to new or unusual situations. It enables the learner to "tie up" new ideas with his past experiences and knowledge. The so-called native intelligence of a person changes very little throughout his life and is not increased by education.

Past Experiences—A person's background of experiences forms the basis for receiving additional knowledge.

Concentration—It is difficult for a person to fix his attention on one idea for very long. Yet he may refocus his attention quickly when his mind wanders. The ability to do this repeatedly over a period of time is known as "power of concentration."

Memory—A person's ability to remember is extremely important in learning. Factors which influence memory are vividness, uniqueness, frequency, and relative importance.

Well-Being—Mental and physical comfort increase one's power of concentration. On the other hand, pain, discomfort, and such emotions as grief, irritation, anger, and worry greatly hinder mental processes. The good training man tries to put his group at ease in a cheerful frame of mind before presenting the lesson.

Self-Confidence—A person learns something better if he thinks he can. Fear of bodily injury, fear of failure or spoiling a job, and fear of criticism or humiliation make learning difficult, if not impossible. The good supervisor never purposely assigns a worker a task beyond his ability to perform, nor says or does anything that would tend to make the learner lose confidence in himself.

Imagination—Imagination is the power to form mental pictures of things not actually present. It helps the craftsman to visualize the finished job before he begins it. A man without imagination can never learn to read a blueprint. (1, pp. 71–72)

MOTIVATION IN THE CLASSROOM

Classroom teaching and pupil learning cannot be simply a matter of presentation of materials to be learned. The introduction of incentives, the development of motives, and the utilization of motives provide opportunities for the teacher *to teach;* their employment in learning makes the chief difference between a teacher and a textbook, or between a good teacher and a poor teacher.

Motivation may be dangerous as well as useful. One such danger lies in the administration of punishment for failure in connection with learning. Before punishment is applied, the teacher should make certain that it is an appropriate incentive in the particular situation. School work, for example, should never be utilized as a means of punishment.

Another danger is that of the learner working for the incentive. Some say that emphasis upon the end (an incentive) rather than on the means (the learning) of attaining the end is undesirable. But so long as the desired learning does take place, the incentive which was used to stimulate the learning is, perhaps, of relatively little significance. That is, so far as the learning itself is concerned, it may matter little that incentive was a material reward, praise, or simply satisfaction in doing a job well; the incentive itself drops out of the immediate picture, and the pattern of behavior which was learned becomes the important feature of the individual's experience.

The danger of competition often has been pointed out. But we can perhaps assume that a certain degree of competition is acceptable in learning situations and that it does not necessarily negate the teachings of democracy. A realistic view might attempt to reconcile rather than exclude competition and cooperation. Competition does affect learning. The institution involving rivalry is conducive to greater effort on the part of the individual. Nevertheless, competition among students in classroom activities probably should be used with discretion. The most acceptable form is undoubtedly self-competition.

The following should be used less frequently and with caution:

1. Informing the pupil of failure in learning
2. Arranging competition
3. Administering reproof or punishment
4. Awarding prizes, honors, etc.

They also say: mere activity on the part of the learner does not insure learning.

Least effective devices in classroom motivation are sarcasm, ridicule, low marks, extra work or a penalty, and reprimands before other students.

The following are motivational devices worthy of frequent use in the classroom:

1. Engendering suitable *mind sets, attitudes, or moods*
2. Acquainting the student with *definite objectives*
3. Informing the pupil of *success* in learning
4. *Testing* achievement frequently
5. *Commending* where commendation is justified
6. Conferring when a *conference* will prove stimulating

The teacher is responsible for creating pupil interest (through community events, extracurricular activities, travel, books, pictures, objects for exhibit, projects, etc.) and for creating student needs (through special projects, laboratory and shop work, etc.).

Other thoughts on motivation in the classroom include:

1. Emphasis on meanings and relationships contributes to the individual's set for learning. Materials lacking in meaning are relatively more difficult to learn than those of which the meaning is understood. Meanings may be developed by basing the learning upon the past experience of the individual, by extending and elaborating (or enriching) the basic principles or facts involved, by taking advantage of the logical relationships of the material in organizing it for presentation, and by pointing out possible applications of the learning. Overviews provide a facilitating set for the learning to follow.
2. Interests, attitudes, and purposes must sometimes be developed, or needs created, as a first step in learning. Needs may be created through the introduction of projects which tap a major interest of the pupil, but, at the same time, demand that he learn a variety of new facts or skills in order to bring the task to completion.
3. Goals and standards to be met function successfully as incentives only when adapted to pupil ability. Care must be taken in making assignments to be certain that the learning required does not exceed the pupil's ability or is not beneath his ability.
4. Definite objectives are necessary if motivation is to be effective (immediate goals more effective: long term and short term). Closeness to the goal is an important factor in learning, and learning behavior is facilitated as the goal is more closely approached.
5. Pupil's interests are important sources of motivation. Participation in meaningful activities furnishes a background for the development of interests.
6. Specific directions and suggestions for learning contribute to the student's set.

7. Rewards and praise may be effective incentives for some students in learning situations. They may lose their value if repeatedly used with the same individual.
8. Punishment is probably of limited value in motivating learning. It must be appropriate.
9. The teacher plays an important part in the motivation picture. Attention should be given by the teacher to personality traits which pupils generally like and others which they generally dislike in teachers. Teachers who are liked and respected themselves contribute directly to the pupil's learning and the pupil seeks to obtain the teacher's approval and to identify himself with the teacher.
10. The whole school situation, often involving extra-school activities, contributes to an individual's behavior; and unless these more inclusive and broader aspects of experience are taken into account in applying motivation to learning, the teacher's efforts may fail to produce the desired effects. (2, pp. 1–3)

The Student Is the Center of Good Teaching

The material, the method, and the facilities are important only as they serve the student.

Study the student as an individual:

Learn about his previous training and experience. What does his background indicate as the starting point for learning the subject to be taught?

Observe his habits of work. These are often cues to other qualities —honesty, loyalty, dependability, etc.

Note his mental strength and limitations—clarity, flexibility, understanding, verbal facility, etc.

Watch his rate of progress. Is he slow, average, rapid?

Assist and encourage the slow.

Raise the level of the average.

Stimulate the rapid to maximum achievement.

Become acquainted with him. Place yourself in his position.

Adjust your teaching to individuals:

Regard your class as a group of individuals, each with his own set of characteristics.

Plan your presentations, discussions, demonstrations, questions, exercises, applications, and tests with the individuals in mind.

Administer the activities of the class with the individual nature of the class in mind.

Remember: The learning is being done by individuals. The class is not a drill team.

General Characteristics of Good Teachers and Good Teaching

The good teacher knows his subject. He never guesses. If he does not know, he finds out. He admits when he is wrong.

The good teacher places a premium on clarity. He is careful to use good English. He adapts his technical terminology to the knowledge of his students. He explains and illustrates all unfamiliar words or those used with special meaning. He insists on the same good standards from his students.

The good teacher is always in control of the class and situation. He avoids familiarity, maintains a dignified attitude with friendliness. He helps his students to control themselves.

The good teacher deals personally with his students without "playing favorites." He is fair and just to all members of his class. He never "picks on" particular persons.

The good teacher makes learning easy. He is efficient in the application of teaching aids, devices, and techniques which speed learning with a minimum of effort. He never tries to make the lesson hard.

Good teaching encourages students to learn for themselves. The teacher does not do what the student can do for himself. He leads students to find solutions; he does not furnish them. The teacher is not a "crutch." He is an assistant. The good teacher sets up situations in which each student is made to depend upon himself.

The good teacher is a stimulator of learning. He primes learning by questions, practice work, cases, and applications. He generates interest. He creates a feeling of need for learning. He sets standards of achievement.

The good teacher is a multiplier of knowledge and skill. His individual power is raised to the equivalent of the number he successfully teaches. His effectiveness is raised to the extent that he raises the effectiveness of his students. Teaching is creative.

TEACHING PERSONALITY IS IMPORTANT

Your personality—the qualities in you that affect your students—are your most important teaching assets.

You will not remake your personality while teaching, but emphasis upon certain traits most effective in teaching will strengthen your work.

The teacher's mental attitude—
Is free from fears.
Accepts criticism.
Is not easily irritated.
Adjusts to the unexpected.
Has control over moods.
Is enthusiastic and happy.
Is loyal to purposes, program, and pupils.

The teacher's personal appearance—
Is well groomed.
Is poised.
Is "wide awake."
Indicates self-confidence.
Reveals good health.

The teacher is cooperative—
Accepts suggestions gracefully.
Admits mistakes without offering an alibi.
Deals fairly with students.
Is sympathetic and understanding.
Takes the lead without dominating.

The teacher is socially adequate—
Is confident.
Shows interest in people.
Puts others at ease.
Is free from irritating voice.
Avoids distracting mannerisms.
Is tolerant of others' views.
Is a good listener.

An occasional personal check against this list will assist you in improving your teaching.

All that a person is counts in teaching.

TEACHING MUST BE PLANNED

As the architect plans a building, as the radio producer plans a program, as the military staff plans a campaign, so *the teacher plans a lesson.*

In order to plan, you must know—
The purpose of the course.
The teaching material of the course.
The characteristics of the students.
The equipment and facilities available.

A lesson should—
Have one main objective.
Be built upon previous lessons.

Advance the study.
Require tangible achievement.
A lesson plan should contain a statement of—
Purpose: The teacher's, the pupils'.
The class activities for reaching the purpose.
What are the students going to do?
The subject matter to be learned.
The technical terms to be explained.
Case or illustrative material to be used.
The time available.
Equipment or other aids to be used.
The questions which motivate and test the learning.
The plan for appraising the achievement.
There are four steps in learning—
The learner's interest is aroused.
The learner receives the lesson.
The learner applies the lesson.
The learner tests his skill or knowledge.
These steps must be repeated until the lesson is learned.
Plan your work and work your plan.

Structure for Planning

The structure for planning to teach is somewhat similar to the process of planning a house. The need is determined by analysis. Analysis of the job provides the information for the total planning of the program.

Program planning begins by the identification of the curriculum content. The curriculum is divided into courses. The courses are subdivided into units. Each unit is composed of two or more lessons. Some of the lessons will be devoted to shop and laboratory activities, while others are planned as related or applied instruction.

To better relate the scope of work encompassed in each unit to the available time, it may be helpful to establish a structure covering the period of time estimated to teach the unit. Now, it is recognized that the rate of progress will depend upon the individual student's ability and motivation. Therefore, such a plan should be viewed as a budget of time rather than a fixed schedule to be rigidly adhered to. A budget of time established and known by both the students and teacher often results in much more effective use of time and learning resources.

A three weeks' planning outline is recommended as minimum advanced planning for each course. This may be done through a simple form as shown on Exhibit 13–1. This planning structure may be limited

EXHIBIT 13–1

THREE WEEKS' PLANNING OUTLINE

COURSE____________________UNIT____________________

Week	MONDAY	TUESDAY	WEDNESDAY	THURSDAY	FRIDAY
1					
2					
3					

to one unit of instruction. It is an excellent way of getting student teachers and regular teachers to think and plan ahead.

Words or phrases in each block provide the keys to lesson planning. Through this approach a coordinated overview is established resulting in better coordination. Gaps or omissions in teaching content are less likely to occur.

This form can easily be modified for a shorter or longer period of time.

After the broad structure for planning of the unit has been designed, it is then wise to translate this into daily or unit lesson plans. These lesson plans are necessary for related lessons or units just as they are essential for shop or laboratory lessons. The lesson plan should serve as a road map. First, the instructor must identify where to go—this is the determination of the objectives. Then, it is necessary to decide what to take along to do the job that has to be done—this constitutes the tools, equipment, and supplies. The next step is to prepare for the trip—this is the preparation. Then the journey is undertaken—this is the presentation. Did the travelers arrive at the destination?—this is the application and the test. Exhibits 13–2, 13–3, 13–4, 13–5, and 13–6 illustrate various types of lesson plan formats and content.

HOW TO MAKE YOUR LESSONS EASY TO FOLLOW

1. Prepare and use a lesson plan to make sure your lessons are accurate and complete.
2. Spend a few minutes reviewing the previous lesson.
3. State the title of the new lesson.
4. Briefly state the objectives so the student will get an immediate overview of what the lesson is about. A blackboard or chart outline is helpful.
5. Don't forget motivation. Explain why each lesson is necessary, *when, where,* and *how* it ties into their work. Use a little salesmanship. It pays off in better student interest, attention, and learning!
6. Vary your method of presentation. Combine demonstrations, questions, discussions, etc., with your lecture. Use the blackboard, charts, films, or other appropriate instructional aids.
7. Stay away from the "straight lecture." It's the poorest method of teaching.
8. Write new terms on the blackboard and explain them. Students can't follow if terms are not clearly understood.
9. Talk to the students—not to the blackboard, equipment, or windows.
10. Make practical applications of the classroom lessons to the shop work, and vice versa.
11. Ask questions frequently—during and after your presentation. You can't be sure your students are following you unless you check.

(text continued on page 217)

EXHIBIT 13–2

Lesson Plan (Manipulative)

Lesson Plan No.______
Use same No. as in course of study

(Insert title of lesson)

Objectives:	List what is to be achieved as a result of the lesson. Should be brief and concise. In some cases one may be sufficient.
Tools, Equipment, and Supplies:	List tools, equipment, and supplies necessary to teach the lesson.
Preparation:	State specifically how attention of students is to be obtained and how interest is to be developed. This step may be any length but is usually three to five minutes duration. This step should lead smoothly into the presentation step.
Presentation:	List each step and key point necessary to perform the operation or operations. Length of presentation approximately 30 to 45 minutes. State steps clearly and concisely in occupational terms. List steps in proper sequence. Number each step consecutively. Key points should be listed with steps where they apply.
Application:	Student is assigned a job from the course of study, involving the operation or operations presented. Supply student with the appropriate Job Sheet or have student prepare the Job Plan Sheet. Effectively motivate students when making job assignment. The job assignment should be such that it can be completed within one week or less. The instructor supervises the work but should not work on the student's job. Provision for individual differences should be made.
Test:	Check quality, time, and techniques. Criteria for checking are to be based on acceptable standards.

EXHIBIT 13–3

Sample Daily or Unit Lesson Plan (General) I

Unit____________________ Topic____________________

Objectives ____________________

Aids and materials ____________________

Preparation ____________________

Presentation
Introduction____________________

Steps	Key Points

Questions
(Summary) ____________________

EXHIBIT 13–4

Sample Daily or Unit Lesson Plan (General) II

Title of Lesson____________________

Objectives of Lesson

1. ____________________
2. ____________________
3. ____________________

Step I. Preparation Steps

A. Preparation by the Teacher

1. Tools, Supplies, Equipment____________________
2. Books, References, Instruction Sheets, Visual Aids____________________

B. Motivation of the Student

1. Introducing the Lesson____________________
2. Association or Connection with the Previous Lesson____________________

Step II. Presentation Step (Underline Methods to Be Used)

Demonstration Lecture Illustration Discussion Experimentation

A. Teaching Points

1.__________	6.__________
2.__________	7.__________
3.__________	8.__________
4.__________	9.__________
5.__________	10.__________

EXHIBIT 13–4 continued

Step III. Application Step

Give Assignments

____

____

____

____

Step IV. Testing Step or Follow-up

List Inspection Points, Questions, or Problems

____

____

____

____

EXHIBIT 13–5

Related Lesson Plan

Course: Basic Electronics

Lesson Title: The Structure of Matter

Objectives: To learn the principles of the electron theory.
To develop its importance to all phases of electronics.

References: None

Materials Needed: Plastic model of atom
Colored chalk

Introduction:

1. State reasons why it is important to study the electron theory.
2. State why it is important to understand the structure of matter.
3. Ask questions: What is matter? What is an electron?
4. Explain that, as we study electronics, we will begin to get the answers to these and other questions.

Presentation (first):

1. Give definition of "matter."
2. Present illustration of the solar planetary system, reviewing the effect of gravitational pull by the sun and the centrifugal force exerted as a planet revolves, causing it to travel in a specific orbit.
3. Illustrate on board the similarity between the structure of the atom and the solar system. Demonstrate with plastic atom model.

EXHIBIT 13–5 continued

4. Define the terms: electron, proton, neutron, and nucleus.
5. Discuss positive and negative charges as related to the proton and the electron and their relationship to the "law of charges." Describe ions.
6. Define "weight" and "mass" as applied to an atom.
7. Define and discuss "elements," using the atomic structures of the elements of hydrogen, helium, and lithium as examples.

Application (first):

1. Ask for questions up to this point in the lesson.
2. Ask questions:

    What is an electron? Neutron? Proton? Atom?

    In what proportion, with respect to protons, do neutrons contribute to the mass and weight of an atom?

    Of what importance are the planetary electrons of an atom with respect to electricity?

    What constitutes a positive ion, a negative ion?

Presentation (second):

1. Give the definitions of a molecule, a compound, and a mixture.
2. Explain how compounds are formed, using examples.
3. Explain "mixtures," giving examples.
4. Illustrate on board and discuss the orbital primary shells and subshells of an atom.
5. Demonstrate with plastic model of the atom.
6. Illustrate on the board an atom of copper, and reasons for ease of electron movement within this element.
7. Explain the binding characteristics of an atom. Three general classes of valence: "ionic," "metallic," and "covalent."

Application (second):

1. Ask for questions covering this part of the lesson.
2. Ask questions:

    What are the essential differences between elements, compounds, and mixtures?

    What is meant by a free electron and what are some substances in which these are found?

    Why are some substances conductors of electricity and others are not?

Summarize:

1. Summarize the whole lesson, touching upon high points in the lesson.
2. Ask for questions covering the whole lesson.

Test:

Distribute the written test for the lesson.

EXHIBIT 13–6

Lesson Plan

Course:	Radio Electronics I
Lesson Title:	How to Read the RETMA Color Code on Resistors
Objectives:	To determine the resistance and tolerance of carbon resistors using the RETMA color code.
References:	Read the following: Hickey and Villines, *Elements of Electronics,* pp. 32–33. Zbar and Schildkraut, *Basic Electricity,* pp. 13–14.
Materials Needed:	1. 10 assorted value resistors for each student. 2. Colored chalk.

Introduction:

You have learned to recognize the resistance symbol, and you have also learned that these resistances have different values.

1. Advantages of the color code in circuit tracing.
2. Universal use of the standard color code.
3. Constant use makes color code a habit.

Presentation:

1. Write the colors and number on the blackboard in proper sequence.

Black	0	Green	5
Brown	1	Blue	6
Red	2	Violet	7
Orange	3	Gray	8
Yellow	4	White	9

2. Draw resistor on the board and indicate 4 bands with colored chalk: red, violet, red, and silver.
3. The band nearest an end is the starting point and denotes the first significant figure.
4. The second color band denotes the second significant figure.
5. The third band denotes the number of zeros to follow.
   a. This is called the multiplier.
6. The fourth band, if there is one, gives the percentage of tolerance.
   a. Gold 5%
   b. Silver 10%
   c. No band 20%
7. If gold or silver is used on the third band:
   a. Gold, multiply by .1
   b. Silver, multiply by .01

EXHIBIT 13–6 continued

8. Read value of the resistor on blackboard.
9. Show four or five examples of four-band resistors.
10. Show several examples of three-band resistors with no fourth band.
11. Show several examples of three-band resistors with gold and silver bands.

Application:

1. Pass out five resistors of different values to students.
2. Have students write the values on paper, adjacent to the resistor.
3. Check students while they work.

Test:

1. Collect resistors used in application, and pass out five more of different values to each student.
2. Have students attach resistors to paper with tape and write the value of each adjacent to the resistor.
3. Write five values of resistors on blackboard and have student write the colors of the bands necessary to obtain that value.
4. Collect the papers and check for errors.

This is very important! Write questions in your lesson plan to ensure asking them at the proper time. Good questions are a great aid to good teaching.

12. Secure student participation. All learning requires activity. Encourage students to ask pertinent questions; stimulate discussion; encourage good note taking and problem solving; utilize friendly competition. Sleepy students aren't participating.
13. Stay on the subject.
14. Reach definite conclusions in all discussions. A point hanging in the air is confusing to students.
15. Summarize main points. Sub-summaries may be advisable.
16. Key questions serve as a good check on understanding as well as a summary of the lesson.

CHECK SHEET FOR LESSON PREPARATION

A. Selecting the Lesson Unit
   1. Does the lesson unit selected deal with *one* topic only?
   2. Is the unit small enough to be covered adequately within the normal interest span of the students, say 20 to 35 minutes?
   3. Is the lesson simple enough to ensure the instruction being effective when the opportunity for its application comes?
   4. Will the lesson be well timed? Does a need exist? Can an application of the new instruction be made immediately or reasonably soon?
   5. Does the lesson confine itself to the material pertinent to the work of the students *at this time?*
   6. Does the lesson have a definite relationship to preceding lessons and lessons that are to follow?

B. Preparing the Lesson Plan
   1. What plan or idea do you have for getting the initial interest of the students?
   2. Has the lesson been built on or related to the previous experiences of the student?
   3. What examples, illustrations, and "for instances" common to all students are you going to utilize in making the importance of the lesson clear?
   4. How have you planned to bring out the significance of this lesson with respect to the trade, technology, or occupation?
   5. Have you anticipated your students asking:

      "Why teach us this lesson now?"
      "What value will this lesson have to us?"
      "Why is it important to know about this?"
      "What does this have to do with our shopwork?"

      Have you planned your introduction in a way that assures answering these questions to their satisfaction?
   6. Have you planned a clear, concise statement of what the lesson is going to be about?

7. Have you carefully analyzed each step or point in the lesson to be presented and written these key items down in the best learning sequence?
8. Have you considered all the possible methods of presenting the teaching points which you have listed? For example:
   (a) Giving a demonstration before the class yourself
   (b) Having another person give a demonstration for you
   (c) Perform the new operation simultaneously with the class
   (d) Explain and illustrate on the board or with charts, pictures, or projection machines
   (e) Perform an experiment for the benefit of the class
9. Do you know of any models, sections, or simplified equipment that will make it easier for the students to grasp your explanation, particularly where some difficulty may be anticipated?
10. Are any charts or diagrams available or could some be made to aid in putting over major points of your lesson?
11. Have you considered the use of moving picture films, slides, or the projection of pictures for this presentation?
12. What drawings could be placed on the board before or during the lesson to visualize the points you wish to get across?
13. Have you tried to call to mind some common everyday experiences that will serve as analogies in explaining unfamiliar things?
14. Have you examined the new instruction material for strange terms and new trade names? Are you planning on writing them on the blackboard and asking the students to copy them for future reference?
15. Do you plan on having the students take notes or make sketches? If so, when and what specific instructions do you intend to give?
16. Have you thought of calling on one or two pupils to perform the operation demonstrated as a means of clarifying the instruction and reviewing the procedure?
17. Have you planned to check your teaching step by step by questioning students on the material being presented?
18. Have you considered the feasibility of having all the students attempt a trial performance of the operation demonstrated as a check of your effectiveness?
19. Could a short paper test be given to determine how well the lesson has been "put over" or would a short oral quiz be sufficient? Could you use both?
20. Are you planning on having the students bring out the main points of the lesson in the summary or are you going to do it?
21. Will the points so brought out be written on the board, copied in notebooks, or just reviewed orally?
22. Have you made a list of the highlights which the summary should include?

23. Have you timed your lesson so that an assignment can be made which provides for an immediate application of the instruction just given?
24. Have you prepared detailed instructions for the students on how they are to proceed in putting the things just learned into practice?
25. Will the students be given a list of reference readings or notebook work as an incentive for further study and accomplishment? If so, have you planned it in a way that:
    (a) Everyone will see the advantage of carrying it out?
    (b) Provision will be made for individual progress?
    (c) Everyone will know exactly what is to be done?
26. Have you a plan for following up the lesson with individual instruction that is systematic, impartial, and certain of overcoming student difficulties growing out of the lesson?

C. Preparing the Materials for the Lesson

1. Have you made a list of all the materials and equipment needed for the lesson and used this as a check sheet in getting things together?
2. Have you checked your equipment to be sure it is in good working order?
3. Have you practiced the operation or refreshed your memory about pertinent facts in order to ensure a smooth performance?
4. Have you prepared the necessary written material such as tests, work sheets, etc.?
5. Have you looked up the reference material to be assigned so that you can give specific assignments referring to exact title, author, and page number?
6. Have you made all necessary arrangements for supplies, tools, and procedures needed to facilitate the application of the lesson following the assignment?
7. What provision have you made or are you planning to make so that all students can see and hear everything without discomfort or distraction?
   For example:
    (a) Large illustrations and models
    (b) Good light on the object being viewed
    (c) Pupils arranged so all can see without crowding
    (d) No unnecessary noise to interfere with hearing
    (e) Placement of visual aids and materials of demonstration where all can see
    (f) Use of color in diagrams, charts, blackboard, drawings, and on models
8. Are you sure you know how to pronounce and spell all new and unusual terms that are expected to be introduced or used?

Plan thoroughly—good planning is the essence of good teaching.

Summary

Effective lesson planning is essential to meaningful teaching and efficient learning. The teacher will be more successful if he (or she):

1. Remembers that the senses through which we learn must be aroused.
2. Plans to use instructional methods that involve a multiple of the learner's senses, i.e., seeing and hearing.
3. Applies knowledge of laws in learning governing habit development, i.e., use, practice, association, etc.

Students differ as to interest, intelligence, past experiences, concentration, memory, well-being and self-confidence.

Motivation is essential but it can be used in destructive as well as constructive manners. Positive motivation helps students to:

1. Establish the correct frame of mind.
2. Work to achieve the performance goals.
3. Enjoy the satisfaction of accomplishment.
4. Acquire individual, or as a member of the group, encouragement to overcome weaknesses and solve personal problems.
5. Recognize the importance of the tasks for present needs and future accomplishments.

The successful teacher is constantly mindful that:

1. The student is the center of good teaching.
2. The teacher's personality influences learning.
3. Teaching must be planned.

Teacher planning of the instructional activities should include a long-range perspective as well as daily or unit lesson plans.

The lesson plan is the road map for the activities to be programmed. It must be rigid enough to provide a step toward an established goal(s) and flexible enough to adjust to the learning needs of individual students. The structure of the lesson plan should include: objectives; tools, equipment, and materials; preparation; presentation; application; and test.

QUESTIONS AND ACTIVITIES

1. Why is planning before teaching important?
2. What senses are most important in learning? Discuss.
3. How can useful habits be most effectively developed?

4. What individual differences should teachers recognize in their students?
5. List and describe constructive motivational methods for vocational teachers.
6. Is sarcasm a good motivational technique? Explain.
7. What is the basis of a student-centered instructional program?
8. Identify the characteristics of a good teacher.
9. What should be included in the lesson plan?
10. Discuss the relative merits of the daily and the unit lesson plans.
11. Why use a planning outline?
12. Develop one or more lesson plans for:
    A. A manipulative lesson
    B. A related lesson

REFERENCES

1. *The Preparation of Occupational Instructors.* Washington, D.C.: Superintendent of Documents, 1965.
2. Purdue University, Division of Education and Applied Psychology, mimeographed, n.d.

Chapter 14 WHEN YOU BEGIN TO TEACH

All learning is individual. Students, whether in groups or singly, learn as individuals. Learning is the responsibility of the student, but, motivation and teaching are the responsibility of the teacher. Teachers create the environment for learning.

Young children usually look forward to going to school. After a few years many students have lost the desire to go to school. This may be the result of what Robert F. Mager describes as aversive consequences. He further describes an aversive consequence as any event that a person will try to avoid. He tells us that while many events may or may not be aversive to a given individual, there are several universal aversives which will act to attentuate an approach probability if they occur concurrently with, or immediately follow, an approach behavior. These universal aversives, Mager says, include fear, pain, humiliation, boredom, and repeated failure. Teachers can control the learning environment to minimize the aversive effect of these undesirable learning experiences, although some will occur as concomitant experiences. Mager illustrates this further, as follows:

> If, every time a student picks up an engineering textbook, he is dropped through a trap door into a pit of alligators, the probability decreases that he will pick up the book in the future. The proba-

> bility of the action being repeated also is attenuated if any one of the other aversive consequences follows the action. . . . To engineer the situation to maximize the strength of an approach tendency, one would eliminate as many aversive consequences as possible, much as one would eliminate as many resistances as possible from a circuit whose current he was intending to maximize.
>
> Another dandy practice for driving students away is that of responding to a student's question with a comment, look, or gesture that causes the student to think less highly of himself or that causes his world to become a little dimmer. . . . When, after having asked a class a question, and after the answer has slipped inadvertently from between the lips of a student so anxious to come into contact with the subject that he couldn't restrain himself, one says pointedly and deliberately: "I . . . didn't call . . . on . . . you." (1, pp. 841–43)

Teaching, then, is the process of enhancing and expediting learning for the student. Every action and activity of the teacher should occur for these purposes.

The teacher controls but the student reacts. The teacher controls the environments for learning. Three main environments are involved. These are the:

Physical environment
Emotional environment
Mental environment

If these environments influence a positive reaction the effect for that student is positive learning. If negative learning results, the student may react violently, seek escape, or accept a self-concept which produces a defeatist attitude.

For some students, learning is more difficult than for others. True, innate abilities, aptitude, and interest vary. But, for many students learning is difficult because of:

Fears and worries—fear of failure; fear of ridicule; worries about family and home problems; worry over money; fear of the instructor.

Discomforts—from standing too long; eye strain; dirty working conditions; dangerous tools and equipment; poor ventilation; poor shop organization.

Boredom—from instructor talking too long at a time; little time to use shop and equipment; instructor not prepared; training aids not used; teacher a "Blow Hard" about himself.

The Teaching Steps

The process of teaching consists of basic steps. These may or may not be completed in one lesson.

From the analysis of a trade, technology, or occupation it is possible to select the skills, knowledge, and attitudes that must be taught. In Chapters 4, 5, and 6 a procedure for organizing and arranging and subdividing this reservoir of teaching material was discussed. Having decided on what is to be taught and when it should be taught, one must then decide on how a lesson should be taught.

It is generally true that there is more than one good method of doing a job. Teaching is not an exception to this rule. There are many different procedures that have been used in teaching. The beginning instructor will be wise to try to master one tried and proven method of teaching before he becomes concerned with other methods.

Vocational teachers have for many years been using a pattern around which any lesson may be outlined and taught. This pattern is made up of the following four steps:

Step I: *Preparation*

The instructor assembles all materials, supplies, and equipment for the lesson. Then he must gain the attention of the student and arouse his interest in the lesson.

Step II: *Presentation*

The instructor demonstrates and explains exactly what is to be learned.

Step III: *Application*

The student performs, individually, the operations to be learned with the aid and assistance of the instructor.

Step IV: *Testing*

The student performs the operations by himself and is checked for efficiency and skill by the instructor.

A beginning instructor will find it desirable to plan each lesson around these four steps, and to follow them closely as he teaches the lesson. At the beginning the new teacher will probably be very conscious of the form of the lesson, but with practice the steps will become second nature. It is essential that instruction in the shop be as informal as possible. Accordingly, the skillful instructor synchronizes these steps into a unified whole. The four steps will now be discussed more fully. See Figure 14–1 for a visual presentation of these four steps.

PREPARATION

Any demonstration, to be interesting and effective, must be smooth and run off like clockwork. To accomplish this the instructor must have everything in readiness before he begins. Usually, it is desirable to run through the demonstration to be certain that it will click when the time comes. The demonstration should always be made with the same type of material and tools that the student will be expected to use. If possible, devise an instructional aid that will assist the student to more clearly

HOW TO GET READY TO INSTRUCT

Have a Timetable
Include how much skill you expect him to have and by what date.

Break Down the job
List important steps.
Pick out key points. (Safety is always a key point.)

Have Everything Ready
Have the right equipment, materials, and supplies.

Have the Workplace Properly Arranged
Arrange workplace just as the worker will be expected to keep it.

Name of
Sponsoring School
Here

Keep This Card Handy

HOW TO INSTRUCT

Step I: Prepare the Worker
Put him at ease.
State the job and find out what he already knows about it.
Get him interested in learning job.
Place in correct position.

Step II: Present the Operation
Demonstrate and explain one IMPORTANT STEP at a time.
Stress each KEY POINT.
Instruct clearly, completely, and patiently, but no more than he can master.

Step III: Application or Tryout Performance
Have him do the job—correct errors.
Have him explain each KEY POINT to you as he does the job again.
Make sure he understands.
Continue until YOU know HE knows.

Step IV: Follow Up or Test
Test to determine if he has learned the skill or information.
Put him on his own. Designate to whom he goes for help.
Check frequently. Encourage questions.
Taper off extra coaching and class follow-ups.

If Worker Hasn't Learned,
The Instructor Hasn't Taught

This illustration is the front and back of a reminder card used by many successful vocational instructors. It tells how to use the four steps when teaching a skill.

Source: The Preparation of Occupational Instructors. Washington, D.C.: Superintendent of Documents, 1966, p. 27.

FIGURE 14–1

THE FOUR–STEP LESSON

understand the new aid. The instructor should make provision for the materials needed by the students, and clearly formulate in his mind the standard of work that will be acceptable from the learner.

The preparation of the lesson is only half completed when the instructor is all ready. The next step is to prepare the student to receive the lesson. There are many ways of doing this, but basically the procedure is the same. It is first necessary to secure the attention of the student and arouse his curiosity and interest in the lesson. Once this is accomplished it will be relatively simple to connect the new lesson with the student's previous experiences. If the instructor has properly performed the preparation of the student, the time will quickly come when the student will ask for additional information. This is the mark of success in step one.

PRESENTATION

The demonstration is now performed and explained. The methods and teaching aids will depend on the lesson, and on the previous experience of the student group. The successful instructor usually goes through the lesson step by step, taking as much time as seems to be desirable. He may pause at various spots to point out details that might escape the attention of the learner or to indicate the common errors. The instructor should ask questions that will encourage the student to think about the job. He should emphasize the safety factors and explain the "why" of the established procedure for the job. Throughout the demonstration the student should be encouraged to ask questions.

A common mistake is to try to do too much at one time. The presentation step should not be too long. It would be more desirable to have two shorter demonstrations than one long drawn out one. The chief concern of the instructor is to show the learner "how" to do the job and to do it safely and efficiently. When the instructor feels that the student understands the lesson and is ready and anxious to try to do the operation alone, step two has accomplished its purpose.

APPLICATION

This is the try-out step. The student is now ready to attempt the job by himself under the supervision and encouragement of the instructor. Everything that the instructor has done up to this time was intended to help bring the student to the place where he can strike out for himself. However, his first efforts are likely to be rough and clumsy and frequently wrong. The student should clearly understand that the instructor is anxious to help him overcome his difficulties in this stage. Equally important, the student and instructor should realize that the student will overcome his difficulties only when he knows how the operation is to be done and then is given the opportunity to do it. In many cases, it will be desirable for the instructor to repeat his demonstration, or some part of it, in order to clear up a point or two. It may be necessary for the learner to repeat the steps of the lesson several times before he acquires the desired skill. When the learner is able to perform the task efficiently and safely, step three has been successfully completed.

TESTING

The only real proof of a good job of teaching is found in the ability of the student to perform the operation efficiently and safely on his own. Accordingly, in the testing step the learner is required to do his best without any assistance. The instructor is thus able to determine exactly the nature of the student's difficulties, if any. Also the learner may satisfy himself that he is really capable of carrying on without help. Thus, a test when used in this way assists both the instructor and student, and indicates whether or not the learner is ready to proceed further.

What Is a Lesson?

"Experience is the basis of all learning." This is a fundamental in trade and technical teaching. The part the teacher plays is that of helping the learner to interpret his past experiences and providing further experiences whereby he may acquire more skill and knowledge. The instructional activities which provide such experiences in a specific and limited topic, skill, or idea, comprise a *lesson*.

A second fundamental is that: "A learner acquires all his knowledge by learning *one* thing at a time." Each skill, principle, or concept that he masters is like a stepping-stone to the next item in the training course. Therefore, each new bit of learning is effective only when it relates back to previous experience. Each stepping-stone then becomes that new experience which is commonly called a lesson.

On the stepping-stone basis, the learner acquires the new experience from four activities:

1. Finding how the assignment fits into his past experience.
2. Watching another and more skilled individual do the operation; that is, seeing it demonstrated.
3. Trying the operation under the supervision of a skilled individual.
4. Practicing until he can do the operation satisfactorily.

CHARACTERISTICS OF A LESSON

When properly presented, a lesson helps the learner much like a road map helps the traveler—it shows him where he is at the moment, where he is going, and puts emphasis on the present lap of the journey. To continue his journey he must go through new experiences—*he must travel the road of learning.* Thus, a satisfactory lesson is designed to meet the needs of the learner. It will meet these needs if it satisfies the following queries:

1. *What relation?* The lesson must show relationship to preceding lessons. Since every new experience is interpreted by the learner through previous experiences, the new materials to be learned must be related to or associated with his former experiences.
2. *What is new?* The lesson must provide something new and definite to be learned. The new experience is the essence of the lesson. If an activity offers no new experience, it presents little opportunity for learning.
3. *How far?* The lesson must be reasonable in scope. It should have sufficient content to challenge the learner's best efforts but must not cover so much ground that he becomes tired or discouraged. The learner should be able to see the successful outcome of his efforts or the completion of his task before interest lags because of fatigue or other interests.

STEPS IN THE PRESENTATION OF A LESSON

1. *Get everyone's attention*
   (a) call attention to a group need
   (b) recall a problem arising in class
   (c) relate an incident which has some bearing on lesson
   (d) read a clipping from a newspaper or periodical pertinent to lesson
   (e) other—seek new and more effective approaches
2. *Introduce the lesson*
   (a) develop a feeling of need for lesson
   (b) connect the lesson with previous work or experience
   (c) review the material prerequisites for the lesson
   (d) other—use variety and initiative in providing variety
3. *Present the new material of the lesson*
   (a) demonstrate
   (b) explain
   (c) show (pictures, charts, models, diagrams, slides, etc.)
   (d) question
   (e) discuss
   (f) experiment
   (g) other—be effective and innovative
4. *Check point by point throughout the lesson; see how you are "getting across."*
   (a) ask questions

(b) call for explanations
(c) provide for student participation
(d) other—observe students' reaction, etc.

5. *Make a final check for thoroughness of the entire lesson*
   (a) use oral questions
   (b) give short written tests
   (c) allow an opportunity for application
   (d) other—find method of assurance of accomplishments
6. *Summarize the lesson just presented*
   (a) state the major points yourself
   (b) write statements on the board
   (c) have students copy summary in their notebooks
   (d) bring out summary statements by questioning
   (e) other—tie the lesson together in a neat package
7. *Give a clear and specific assignment; provide for an application of the lesson just presented*
   (a) assign notebook work
   (b) assign shopwork on a job or project
   (c) assign reference work for further study
   (d) assign problems for solution in or out of school
   (e) assign observations and reports
   (f) other—induce activity that results in positive applications of learning

USE OF QUESTIONS

The dictionary defines the word "Question" in this way . . . to ask, to investigate, to seek, to inquire and to learn. It is hard to believe that such a simple thing as a spoken question is one of the keys to human conduct, achievement, cooperation, and education. The power of the question lies in its requirement of an answer and in this way we learn. A judicious use of questions is one of the most desirable and effective learning tools, when it is coupled with an ability to listen and get the other fellow's viewpoint. We learn best through seeing. The next best way, of course, is through hearing and with hearing go questions. This is practical learning to the best degree. The following outline identifies briefly the main concepts and uses of questions.

I. Four Basic Types of Questions are:
   A. Overhead
   B. Reverse
   C. Relay
   D. Direct

II. Definition of Each
   A. Overhead—a question asked to the class with no particular student specified to answer

B. Reverse—the question is answered with a question
C. Relay—the instructor does not answer himself but relays the question to some other student
D. Direct—the instructor asks, pauses, and then indicates who is to answer

III. How to Use Each Basic Question
A. Overhead—to start a discussion or to broaden it
B. Reverse—the student's question would be answered with a question. Or, if someone in the class insists on talking on a subject irrelevant to the topic, use this type of question . . . "Will you please write it down and we will discuss it later?"
C. Relay—if you ever get "put on the spot" use it. It is effective and usually stops students from trying to put the instructor "on the spot."
D. Direct—ask, pause, and then indicate who is to answer.

IV. The Purpose of Using Questions
A. As a thinking device to
1. Get attention
2. Recall a fact
3. Create interest
4. Check if student understands what he is to do
5. Make student reason from cause to effect
6. Correct misunderstanding
7. Analyze steps in doing a job
8. Test . . . instructor, did he do a good job . . . student, did he grasp the subject
9. Find out what student actually knows
10. Summarize

V. Features of a Good Question
A. Should be clean-cut
B. Should be understood by the student
C. Should be asked in such a way that the student must think before he can answer
1. Question
2. Pause
3. Indicate who is to answer
4. Answer

VI. How to Use Questions
A. A judicious use of questions is one of the most desirable and effective learning tools that an instructor can use, when it is coupled with an ability to listen and get the student's ideas.
B. Never try to do too much yourself. Others want to talk

too. Let others do the work, by asking questions. Keep the class thinking; use relay, reverse, or direct questions.

C. By a slight rephrasing or clarification of the question asked, the instructor can shape or guide the discussion in the direction that he wants it to go.

D. Questions should be reworded or rephrased when they are vague, ambiguous, or confusing in nature.

E. Use shop questions in the related classes.

F. With a large class, the instructor might not know the names of all of the students. Two methods help:
   1. Seat students alphabetically
   2. Use a seating chart

G. To hold attention of students.

H. To create interest.

I. Ask a student more than one question a period if possible.

J. Fit the question to the student:
   1. Tough question to brighter students
   2. Easier question to backward students

K. Use a provocative question to encourage discussion.

L. Use a third-person question if you think the class is prejudiced.

VII. Things to Avoid in Using Questions

A. Arousing antagonism
B. Personal questions
C. Building up the instructor and belittling the student
D. Putting someone "on the spot"
E. Sarcastic questions
F. Ambiguous questions
G. Chorus recitation
H. Rotation answering
I. Repeating questions
J. Repeating answers
K. Embarrassing a student

It has been said that skill in the art of questioning is the basis of all good teaching. Good questioning arouses interest, stimulates thinking, and leads to a real mastery of the subject. The efficiency of instruction can be judged, therefore, by the kind of questions asked and by the care in which they are worded. The instructor who succeeds is the one who has developed, at least, a fair mastery of the art of questioning.

Do not make the mistake of spending most of your time asking questions to find out whether or not the students have learned the techniques or skill. In the first place, the teacher cannot get around to any one student often enough to find out very much about what he knows. In the second place, the questions that are asked of different students are of unequal difficulty, and therefore not good for comparing members of the group.

It is impossible to judge a man's knowledge accurately by a few questions. If testing by questions must be done to make sure the day's lesson is learned, a written test is much better than a recitation. In five or ten minutes the whole class may be tested on each of twenty or thirty short answer questions based on main points in the day's lesson, thus providing a better and more reliable picture of the knowledge of all students.

HINTS ON QUESTIONING

1. The questioner should be sympathetic. Encourage the student; do not make him feel that he is in the witness box undergoing cross-examination.
2. In group instruction address your questions to the entire class.
3. In group instruction, state the question before calling the name of the individual who is to answer. In this way, the whole class will be ready to answer.
4. Give time to think between stating the question and calling for the answer.
5. Vary the order of calling on individuals. Call on all students, not just the "brighter" ones. Ask the abler students the more difficult questions.
6. Encourage students to ask questions of you and of the class.
7. Congratulate a student upon an excellent answer to a question.

HOW TO USE THE CHALKBOARD

A teaching skill as important as questioning is the effective use of the chalkboard. Most classrooms and many shops and laboratories provide chalkboards. As the instructor makes the presentation, the chalkboard can be a very useful aid in presenting a more meaningful explanation by developing an appropriate sketch, diagram, outline, or by writing key words as presented in the development of the lesson. Several other excellent rules follow relative to the ways to use the chalkboard effectively.

Chalkboard work should be simple and brief. Copying lengthy outlines or lists of subject matter is a waste of time to instructor and trainee. If it is important for the trainee to have a copy of this material, it should be duplicated and distributed.

The chalkboard is similar to a store window. Everyone knows that an overcrowded, dirty, and untidy window display has little "stopping" value as compared to one that is clean and neat and displays a few well-chosen items.

The following rules for using the chalkboard should definitely increase its effectiveness as a visual aid:

1. Don't crowd the chalkboard. A few important points make a vivid impression.
2. Make the material simple. Brief, concise statements are more effective than lengthy ones.
3. Plan chalkboards ahead. Keep the layouts in your training plan folder.
4. Gather everything you need for the chalkboard before the group meets—chalk, ruler, eraser, and other items.
5. Check lighting. Avoid chalkboard glare. Sometimes it will be necessary to lower a shade and turn on the room light.
6. Use color for emphasis. Chrome yellow and pale green chalk are more effective than white chalk.
7. Print all captions and drawings on a large scale. The material must be clearly visible to each trainee.
8. Erase all unrelated material. Other work on the chalkboard distracts attention. Use a board eraser or cloth, and not your fingers.
9. Keep the chalkboard clean. A dirty chalkboard has the same effect as a dirty window.
10. Prepare complicated chalkboard layouts before the group meets. (2, p. 123)

EVALUATION OF INSTRUCTION

Evaluation of instruction may take the form of self-evaluation or it may be evaluation by a supervising teacher or other administrative personnel.

The main purpose of evaluation is to improve the instructional process. It is also necessary relative to questions of promotion or tenure of a teacher, or in the case of a student teacher to determine the grade earned.

Several forms may be used as an aid in making such an evaluation. Exhibits 14–1, 14–2, and 14–3 illustrate some that are used for this purpose.

Questions Relative to Teaching Vocational Subjects

At this point in progressing through this textbook, it is well to pause and reflect on some vital questions. Some are review questions, while others will be covered in later chapters.

EXHIBIT 14–1

COMMENT SHEET ON EFFECTIVENESS OF TEACHING

____

Taught by______________________________Date______________

Comments by____

STRONG POINTS	POINTS THAT NEED STRENGTHENING

'XHIBIT 14–2

Vocational Teacher Training Summary Sheet for Lesson Evaluation

Taught by____________________ Rated by____________________

Date____________________

	U	P	F	G	VG	PFT
I. *Teaching*						
A. Success in getting class attention						
B. Connection with previous work						
C. Motivation—creating a need						
D. Statement of the lesson						
E. Method of presentation						
F. Ability to explain						
G. Extent and quality of questioning						
H. Distribution of questions						
I. Student participation (verbal)						
J. Extent and ingenuity of checking						
K. Quality of summary or review						
L. Quality of the assignment						
M. Use of visual aids and devices						
N. Class management and control						
O. Ability to stick to the topic						
P. Naturalness of the presentation						
Q. Appropriateness of lesson unit						
II. *Other*						
A. Quality of voice						
B. English—manner of speaking						
C. Appearance—dress						
D. Poise—general bearing—movements						
E. Attitude toward class						

CODE: U—unsatisfactory P—poor F—fair G—good VG—very good PFT—perfect for time

EXHIBIT 14–3

STUDENT TEACHER CHECK SHEET FOR SUPERVISING TEACHERS

Student Teacher) (School)

(Visited by) (Time) (Date)

Suggested Approach:

1. Identify the items on the CHECK SHEET II that apply. Using the numbers as a guide, with the code indicated determine the most objective indicator of performance or status. Establish the profile by connecting the determined points. If desired, the points may be placed between the numbered values. The number 3 indicates average performance.
2. Indicate under "comments" findings that need additional explanation. Use this space for other important factors not listed.
3. The space "recommendations" is intended for summary statements indicating suggestions for strengthening the teaching-learning situation.

Comments:

Recommendations:

CHECK SHEET II

KEY
0—n.a. 3—average
1—very poor 4—above avg.
2—below avg. 5—excellent

ITEM	PROFILE 0	1	2	3	4	5
A. Instructor Planning						
1. Use of course outline						
2. Use of daily lesson plan						
3. Use of course materials, as information sheets						
4. Provision for student evaluation						
B. Presentation Methods and Techniques						
1. Use of demonstrations						

EXHIBIT 14–3 continued

	0	1	2	3	4	5
2. Use of well-planned and stimulating lectures						
3. Use of class discussion						
4. Employs the problem-solving approach						
5. Use of visual aids						
6. Use of chalkboard						
7. Asks stimulating questions						
8. Uses text and other reference materials						
9. Motivates and stimulates students						
10. Relates “known” to the “unknown” content						
11. Employs evaluation, check, or summary						
12. Regulates student participation fairly						
13. Talks clearly and audibly						
14. Use of group instruction method						
15. Use of individual instruction methods						
16. Demonstrates processes and equipment						
17. Holds the attention of the class						
18. Defines and explains terms used						
C. Classroom Organization and Management						
1. Housekeeping						
2. Materials and equipment ready for teaching						
3. Starts and stops on time						
4. Follows through on teaching planned topics						
5. Progress chart maintained						
6. Student-teacher relationships						
D. Physical Organization of Classroom						
1. Ventilation						
2. Lighting						
3. Heating or cooling						
4. Size						
5. Arrangement of furnishings						
6. Adequate chalkboard						
7. Adequate bulletin board						
8. Seating arrangement						
E. Student Preparation and Attitude						
1. Punctuality						
2. Interest						
3. Evidence of preparation						
4. Participation						
5. Cooperation and self-control						

Others require reading, research, and individual reflection. With positive responses to these 100 questions are found the answers to good teaching.

1. What should a teacher do when he is unable to answer a pupil's question? Why?
2. From the standpoint of psychology should terms, definitions, and explanations precede or follow concrete experiences (e.g., a demonstration or trial) with the situations to which they apply? Why?
3. Should pupils always understand the aim or issue before beginning to learn a particular method or process?
4. What is the difference between an inductive and deductive lesson?
5. When should assignments be given?
6. What disadvantages do the true–false type questions have?
7. When is the appropriate time to present a lesson?
8. What different methods should a trade and technical teacher have at his command for effective instruction?
9. Distinguish between interest and attention.
10. How important is motivation in teaching?
11. What specific techniques are effective in creating motivation?
12. What characterizes a good question when used as a teaching device?
13. Enumerate the conditions essential for a good demonstration.
14. By what teaching techniques may provisions be made for individual differences?
15. What determines how long a lesson presentation should take?
16. Differentiate between the use of tests for instructional purposes and tests for administrative purposes.
17. With what books on "Methods of Teaching" are you familiar?
18. Under what conditions would an exercise become effective as a teaching technique?
19. Should the procedure of distributing scores and interpreting them into grades or marks be based on the general performance of the class? Why?
20. Lesson plans are frequently required by administrators. What justification is there from the teacher's point of view?
21. What conditions must prevail if safety instruction is to become meaningful and effective?
22. Are lesson plans worth saving and, if so, what use would you make of them?
23. What bearing does experience have on learning?
24. Students are frequently asked to take notes, write reports, and take down information dictated by the teacher. Can you justify this procedure educationally?
25. Upon what basis should marks in the shop and/or laboratory be based?

26. It has been said that good assignments are fundamental to good teaching. What do you consider a good assignment?
27. How can you justify the use of written tests in the teaching of your shop or laboratory subject?
28. What values do you attach to instructional sheets as teaching aids?
29. In what way will poor methods of teaching contribute to discipline problems?
30. "If a teacher is to secure vocational responses from his pupils the teaching environment must be vocational." If this is true, what are its implications for the teaching of your trade or technology?
31. Under what conditions may repetition be used as an effective learning technique?
32. Enumerate the steps *in their sequence* which are usually evident in a good lesson presentation.
33. It has been said that "the time to teach a boy anything is when he feels a need for it." Point out how this will affect your method of teaching.
34. Describe how a testing program would be used over a term's work and justify its use.
35. What are visual aids and why does one hear so much about them in teaching?
36. Why is it so essential for a successful teacher to "talk the boy's language" and use examples and illustrations within the boy's own experience?
37. We are born neither honest nor dishonest. How did we learn to be honest? What significance does this have for developing desirable personality traits?
38. What is the distinction between "subjective" and "objective" in the matter of tests?
39. In what way does the teaching in evening school differ from teaching in the regular all-day vocational school?
40. Should a shop teacher work out a carefully detailed course of study for each course that he is asked to teach?
41. Is it advisable to assign definite work stations to each pupil in the class? Why?
42. Why is telling and talking too much not considered good teaching?
43. If a trade or technical teacher should maintain close industry contacts, how may it be done most effectively?
44. Assuming that a pupil is capable of only mediocre performance of some trade or technical skills, what are the conditions necessary to bring about a marked improvement?
45. What are some of the principal conditions involved if successful individual instructions were to be introduced exclusively in our school system?
46. What are the essential differences between the methods of training for skill and those for thought and reasoning?

47. What bearing does mental ability have on probable trade or technical success?
48. What objections are usually raised to the "trial and error" method of learning trade and technical skills?
49. To what extent does "satisfaction to the pupil" become a factor in the learning process?
50. Do you know that environmental factors influence individual learning ability? What are some?
51. What is the value of printed instructions as compared with individual oral instruction?
52. What difficulties arise from giving a learner too much assistance?
53. What significance does the following statement have for teaching: "A thing should be taught in the way it is to be used later"?
54. Do you know the meaning of "reliability," "validity," and "objectivity" when applied to tests and testing?
55. When do you think it advisable to call upon students to make shop and/or lab demonstrations?
56. What place do textbooks, reference books, periodicals, and pamphlets have in the trade and technical school shop?
57. What are the factors that make for the most effective teaching of related trade information?
58. Why are examples, analogies, illustrations, and "for instances" so helpful in explaining and describing some point?
59. What specific visual aids are of prime importance in teaching trade and technical subjects?
60. If you were preparing a lecture, what procedure would you use to help ensure its being an educational success?
61. What are the psychological reasons advanced to support or discourage note taking?
62. As a teacher, how can you determine to what extent your lesson has been a success or failure?
63. In what ways are the teacher's effectiveness and the student's achievement related?
64. What plan can be employed to assure that individual instruction is impartial, comprehensive, and efficient?
65. Should teachers of vocational and technical subjects conform to the accepted practices in marking, grading, and distributing marks that are now used in the teaching of other subjects?
66. "Thought provoking" is frequently referred to as a criterion of good instruction. Why?
67. How should the element of speed be handled in relationship to quality of work in shop and laboratory classes?
68. To what extent should the vocational and technical teacher give consideration to the normal curve of distribution in assigning student's marks?
69. What is exploitation? Does production work necessarily imply it?

70. Distinguish between experimentation and trial and error.
71. To what extent should notebooks be advocated for students enrolled in trade and technical subjects?
72. What are the factors which should determine the length of time given over to the actual presentation of a lesson?
73. What distinction should be made between habit and skill?
74. On what educational basis can the use of a monitorial or personnel organization in trade and technical classes be justified?
75. What significance does the "curve of learning" and the "curve of forgetting" have for shop and laboratory teachers?
76. Motivation is sometimes classified into two types referred to by various names, e.g., primary and secondary or intrinsic and extrinsic. Distinguish between the two types.
77. In what specific way has psychology contributed to the improvement of trade and technical teaching?
78. How do the methods employed in teaching trade and technical subjects differ from the methods used in academic subjects?
79. What is the difference between teaching and learning, teaching and telling, and teaching and showing?
80. What would be the disadvantage of trying to learn a trade or technical occupation through a correspondence course?
81. "Practice makes perfect" is an old maxim of learning. To what extent is it true?
82. If success encourages and failure disheartens, should teachers strive to avoid all student failures?
83. If teaching is helping another to learn, what are the things a teacher can do to be of such help?
84. What use might be made of "pre-tests" in trade and technical education? Should these be used?
85. "To question well is to teach well." Explain.
86. Why is it imperative that a trade or technical school exemplify model tool storage and maintenance?
87. What value does organized and planned instruction have over the "pick-up" method?
88. One hears much of the "project method" in general education. How does this method fit into trade and technical education?
89. Some lessons are more successfully taught when each pupil performs simultaneously and step by step with the teacher. When is this technique desirable and why does it work better than others in these cases?
90. "All effective instruction should follow a pattern." Do you know what this basic pattern is and how it relates to present teaching methods?
91. Early psychologists advanced three laws of learning. Can you explain what they are and indicate what reservations are held today concerning them?

92. A blackboard is often considered a teacher's best friend. What are some of the requirements for using the blackboard effectively?
93. What distinction exists between the informational and the developmental approach in lesson presentation?
94. Related and applied information is occasionally segregated from the trade or technical performance and taught separately, usually for administrative purposes. Is this good practice educationally? What educational principle supports your point of view?
95. How do appearance, voice, poise, and emotional stability enter into teaching as factors for success or failure?
96. What rules or principles can you lay down as guides to proper habit formation?
97. An authority on public speaking has said that the formula for a good speech can be written on your thumbnail. It is *"Ho hum, Why bring that up, For instance,* and *So what?"* How does this apply to the presentation of a lesson?
98. Gestures, mannerisms, and overworked phrases in speech often become detrimental to the teacher. What are some of these undesirable speech habits and at what point do they become objectionable?
99. "Be specific" is often cited in education as a guiding principle. Explain why this is true.
100. Some teachers seem to be continually on the defense in their shops or labs. That is, the students are gathered around them, usually at their desk pressing them with questions, work to be examined, or jobs to be approved. What is wrong with this situation and what would you suggest teachers do about it?

ROLE OF TEACHERS AND THEIR IMPACT ON STUDENTS

What is the contribution of the teacher to the student's learning process? What impact does the teacher have upon the students? How relevant and meaningful is the work of the school? These questions and many similar ones need to be constantly asked. Then action must be taken to produce vibrant learning and meaningful education for each student. This is illustrated in the brief story that follows:

I TAUGHT THEM ALL

I have taught in high school for ten years. During that time I have given assignments, among others, to a murderer, an evangelist, a pugilist, a thief, and an imbecile.

The murderer was a quiet little boy who sat on the front seat and regarded me with pale blue eyes; the evangelist, easily the most popular boy in school, had the lead in the junior play; the pugilist lounged by the window and let loose at intervals a raucous laugh that startled even the geraniums; the thief was a gay hearted

Lothario with a song on his lips; and the imbecile, a soft-eyed little animal seeking the shadows.

The murderer awaits death in the state penitentiary; the evangelist has lain a year now in the village churchyard; the pugilist lost an eye in a brawl in Hongkong; the thief, by standing on tiptoe, can see the windows of my room from the county jail; and the once gentle-eyed little moron beats his head against a padded wall in the state asylum.

All of these pupils once sat in my room, sat and looked at me gravely across worn, brown desks. I must have been a great help to these pupils—I taught them the rhyming scheme of the Elizabethan sonnet and how to diagram a complex sentence. (3)

Summary

When you begin to teach remember to try to apply the concept that:

1. Learning can only be accomplished by each individual.
2. Aversive consequences discourage, retard, or eliminate learning for the individual affected. These aversive consequences include: pain, fear, humiliation, frustration, boredom, and repeated failures.
3. The fundamental process of teaching consists of:
   a. Preparation
   b. Presentation
   c. Application
   d. Test
4. Each lesson is a road map or a part of the total road map for the learner traveling toward the vocational objectives of his choice.
5. Effective use of questions improves learning and expedites teaching. Master the application of the four main types:
   a. Overhead
   b. Reverse
   c. Relay
   d. Direct
6. The chalkboard is like a store window. It can be a very effective aid in achieving the objectives identified within the lesson. Learn to master its use in teaching.
7. Evaluation follows teaching. It may be consciously or indirectly done by supervisors, other teachers, or the students. Regular teachers as well as student teachers need to be aware of the checking items and implications for promotions, tenure, or

grades. Self-evaluation is a step toward self-improvement. Use the 100 questions to check your concepts and to aid in reviewing your applications in actual vocational and technical teaching situations. Thus, take whatever steps are necessary to do better.

QUESTIONS AND ACTIVITIES

1. Who is responsible for the student's learning? Explain.
2. What is meant by "aversive consequences"? Discuss.
3. What makes learning difficult? Why?
4. Name and describe each of the steps of the four-step process.
5. What is involved in the preparation step?
6. Explain:
   a. "Experience is the basis of all learning."
   b. "A learner learns one thing at a time."
7. What are the characteristics of an effective lesson?
8. What are the advantages and limitations of each of the following:
   a. Overhead questions
   b. Reverse questions
   c. Relay questions
   d. Direct questions.
9. Explain how to use questions most effectively as a teaching device.
10. What principles should be applied in using the chalkboard?
11. What are the implications for the improvement of teaching of:
    a. Self-evaluation
    b. Evaluation by other teachers or supervisors
12. Should students be asked to evaluate the teacher *or* to evaluate his (or her) learning opportunities? Explain.
13. How can teachers produce a greater positive impact on students? Discuss.
14. Teach one or more short lessons using the lesson plan(s) developed in a previous chapter. Ask your associates to evaluate your effectiveness.
15. Evaluate the effectiveness of the short lessons taught by your classmates. Suggest improvements.
16. Evaluate the opportunities for learning in the other courses which you are presently taking. Suggest positive criticism which should lead to increased learning opportunities.

REFERENCES

1. Mager, Robert F., "Engineering—of Behavior?" *Engineering Education,* March, 1969, pp. 841–43.

2. *The Preparation of Occupational Instructors.* Washington, D.C.: Superintendent of Documents, 1966.

3. White, Naomi John, The Clearing House, Fairleigh-Dickinson University, November, 1937.

Part IV EVALUATING PERFORMANCE

Chapter 15 EVALUATING PERFORMANCE OF STUDENTS

A fundamental goal of all vocational and technical education is preparation for employment. Determination of whether or not the student is ready for employment is a function of performance evaluation. Several methods and a variety of instruments may be employed in determining the achievement of the students.

Evaluation of performance of the students is directly related to the goals established for the educational program.

Purposes of Evaluation

While the major purpose of evaluation of performance is to determine whether or not the student has achieved the goals established for the curriculum, course, unit, and lesson to the degree required on the job for which he or she is being trained, there are several other significant purposes of evaluation.

Evaluation of performance integrates the vital elements of the total instructional system. The system begins with the identification of need, expands into the development of educational objectives, adds instructional materials and processes, and terminates in evaluation of performance of the students. Evaluation provides the essential feedback for refinement and improvement of the total educational process.

Evaluation has been described as process evaluation and/or product evaluation. Process evaluation is concerned with the effects and impacts of the learning–teaching experience during the time the student is in school. Product evaluation reflects the success of the student after the completion of the formal period of education or training and indicates the success of the student on the job. While very important, product evaluation requires a period of time for completion after the student has completed the educational program. It might be said that not enough of this type of evaluation has been carried on over a long period such as five to ten years. To ascertain the success of the student at the time of formal preparation, the process type of evaluation is used.

Evaluation of learning is also a motivational device. "Nothing succeeds like success!" Evaluation helps the students to know the degree to which they have achieved the goals established for the lessons or the course.

Evaluation is fundamental to advancement in the educational program. The determination of the students' progress indicates the need for reinforcement or for reteaching, or the readiness for advancement to the next unit of instruction. Likewise, promotion or completion of the course is usually based on some form of evaluation of performance.

In most courses and schools, performance evaluation is required as a basis for assignment of grades. Grades are required for administrative purposes both within the school and due to organizational policies and structures of the state. These are also important for transfer between schools and for accreditation purposes.

Philosophy, Objectives, and Performance Evaluation

The philosophy of vocational and technical education of teacher and administrative personnel, as well as the policies of the administrative board, must be apparent in the objectives, instructional program, and evaluation.

If the instructional program is the outgrowth of validated need studies, the foundation has been laid for determination of effective performance goals. If the performance goals are defined at the operation or task level, these are easily translated into test items for evaluation of individual performance. However, if only broad general curriculum or course objectives are determined, it is very difficult to know when these have been achieved.

EVALUATION ON AN ABSOLUTE OR RELATIVE BASIS

The demands of vocational and technical education require that students who have completed the program of instruction be capable of performing entry-level jobs. This means that the standards of the job must be considered in making evaluations of competency. This is a very different concept from that most frequently applied in academic courses where a relative performance level is applied.

An absolute criterion required for entry-level jobs may require a fixed percentage of achievement (such as 90 percent) to be employable. Translated into terms of instructor objective for the course, the criterion may be established that 90 percent of the students shall achieve successfully on 90 percent of the subject matter taught. This is very different than the situation in which the achievement of the student is compared in a relative sense with that of other students in the class, as is frequently done on the basis of the bell-shaped curve.

While the basis for assigning grades in academic courses may correctly be illustrated by the normal curve as shown in Figure 15–1 (see also Chapter 21 for further information), the curve skewed to the right is more likely a realistic standard for determining whether or not the performance of the students has been satisfactory in vocational courses. Entry-level employment may demand a high minimum standard of performance by the individual if he is to be retained. The vocational courses in school must reflect these standards realistically.

The philosophy of the standard distribution with 7 percent failures, 7 percent A's, 24 percent D's, 24 percent B's, and 38 percent C's is not suitable for evaluation of such employment objectives. The students must be brought to the level of competency required for employment, otherwise failure is the result. Therefore, the normal curve is skewed to the right resulting in a larger than normal number of high grades and a considerably less than normal number of low grades if the educational program has been effective.

The significance of the concept was emphasized in an article in *Phi Delta Kappan* by Leon M. Lessinger and Dwight H. Allen. This article contained the following statements:

> The fact is that educators have failed to develop performance criteria for measuring the effectiveness of instructional programs, and many programs have been funded and are now under way which at no point describe what students are expected to gain from their educational experiences. What such programs should (but do not) include are agreed-upon standards which would demonstrate that, as a result of the instruction, a change in student behavior has occurred, a change in the direction desired which meets cri-

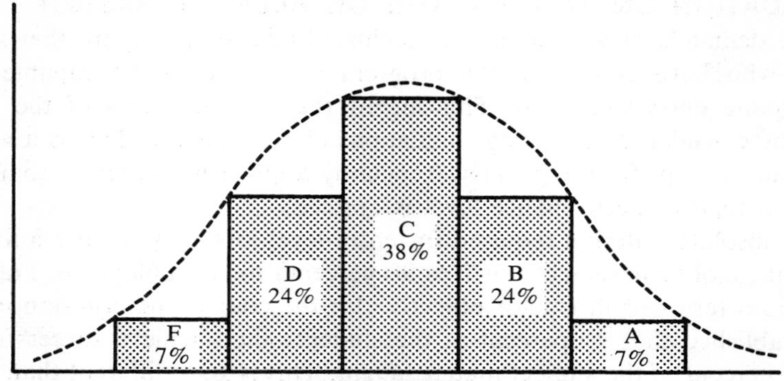

Grade distribution based on normal curve

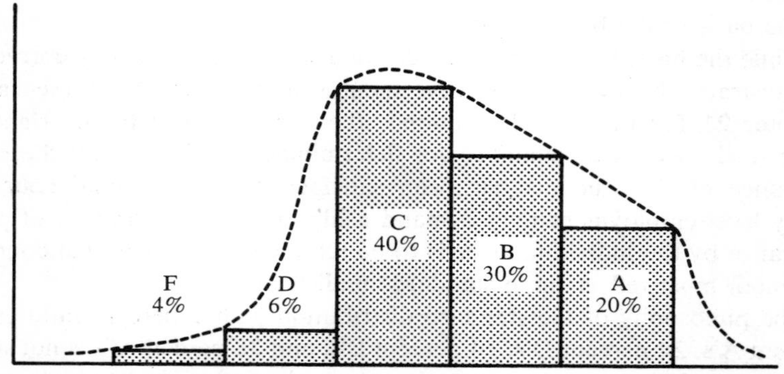

Grade distribution based on curve skewed to the right

FIGURE 15–1

COMPARISON OF GRADE DISTRIBUTIONS BASED ON NORMAL CURVE AND CURVE SKEWED TO THE RIGHT

teria formulated before instruction began. These criteria would clearly specify what it is the student is expected to do, the circumstances under which he should be able to do it, and the degree of accuracy expected.

Instead of brave, vague promises to provide students with "an opportunity to learn to communicate effectively," instructional program objectives should be stated in terms as specific as those in the following example:

> Given three days and the resources of the library, the student completing this program will then be able to write a 300 to 500 word set of specifications for constructing a model airplane that any wood-shop student could follow and build to specifications. (1, pp. 136–37)

With performance criteria expressed in such terms it is simple to establish a valid basis for the evaluation of successful performance.

A prospective employer is more concerned with an applicant's ability to perform against the criterion of the job requirement than he is to know how this student compares on a relative basis with other individuals who might have been in the class at that time.

In related mathematics and science, for instance, the students must know how to apply the principles to the types of problems that must be solved. This is true likewise of related drawing and technical writing—the related information is vital to performance of some part of the required tasks.

MEASUREMENT, EVALUATION, AND ASSESSMENT

Frequently, we hear in relation to student achievement the words measurement, or evaluation, or assessment. Implicit in these terms are degrees of precision and accuracy.

Measurement is the determination of how much. In mathematics it is frequently stated that measured values are relative values; whereas, counted values are absolute. "This engine has eight cylinders" is an example of a counted value. "The page is six inches wide" can be described as a measured value. The accuracy of the measurement depends upon the refinement of the objective and the capability of the individual in reading the instrument as well as the refinement of the measuring instrument. What is an inch? Some would define an inch as $\frac{1}{12}$ of a foot or $\frac{1}{36}$ of a yard. But a more precise definition of an inch is 39,450 wavelengths of cadmium red. Measurement, then, implies a precise determination of a quantitative value.

Evaluation is a broader term which is more dependent upon subjective judgment. It is a more inclusive term which includes the concepts of tests and measures, plus determination of performance on the job through such devices as observation, inspection of finished product, and study of the work in process.

On a hierarchical scale, assessment may be considered as an estimated value applied to the achievement of the students in relation to desired goals.

Relatively speaking, measurement may be considered the most precise with assessment the least precise. Evaluation combines some of the features of each.

The Evaluative Process

The evaluative process will be determined in accordance with the goals that have been set. The conditions under which the determination of achievement is to be made will also affect the process used in evaluation.

The same method of evaluation may not be suitable for the determination of the achievement of skills as for evaluation of knowledges required, or of habits as compared with knowledges, or attitudes as compared with skills.

Frequently, vocational and technical teachers have tried to determine the degree to which students have developed skills by the use of paper-and-pencil tests. These tests are helpful in determining achievement of related knowledges but, frequently, are very inadequate to identify the degree of success of the student in the achievement of skills. The process used in evaluation must be selected in accordance with the needs to be determined.

SECURING INFORMATION FOR EVALUATION

Some of the common methods used to secure information about the progress of the student are:

1. Paper-and-pencil tests
   a. Objective—new type of tests
   b. Subjective—essay tests
2. Oral questions
3. Object tests
4. Performance tests
5. Observation of students at work
6. Inspection of completed job or project

Insight into students' achievement requires determination of the effectiveness of learning of related information. Frequently prospective employers are concerned about the habits and attitudes of the individual seeking employment. The teacher has a responsibility to provide as valid and reliable information as possible to the employer in all areas requiring evaluation of students' progress.

ACHIEVEMENT TESTS

Many different types of tests have been developed. Some tests are designed to measure scholastic aptitude, others special aptitude, still others interest; while another group is intended to identify the characteristics of personality. The focus of tests in this book will be on achievement.

Achievement tests are of two types: either teacher-made or standardized tests.

The standardized test has been refined and validated; whereas, this usually has not been done with the teacher-made test. While the items that are used in the two types of test may appear similar, those in the standardized tests usually have been selected after much refinement. This is followed by a period of statistical validation using item analysis. Then the test is given to a large number of students and norms are developed.

CHARACTERISTICS OF GOOD TESTS

Effective vocational and technical education results when the education is relevant to the needs of the job and the interests of the students. A good test emphasizes the applications of skills and knowledges to the job. "Cold-storage materials" or abstract concepts are much less important in vocational education than applications—test for applications. The ability to use and apply the things taught are the important outcomes of practical instruction. The needs of the student on an entry-level job are the focus of testing and evaluation just as it is good teaching in vocational and technical education.

It is a good policy relative to performance evaluation to give comprehensive tests frequently that test the students' ability to apply what they have learned.

The basic characteristics of good achievement tests have been well documented by authorities such as Micheels and Karnes (2). Let us examine some of these characteristics.

Validity. A good test in order to be valid must measure that which it is supposed to measure. If performance goals or behavioral objectives have been properly determined, the terminal behavior desired has been identified, defined, and the criteria of acceptable performance specified. It is then a simple matter to translate this into a test item for performance evaluation.

Reliability. A good test is consistent or reliable. It measures in the same manner each time that it is used. Reliability of a test can be determined by using

1. two equivalent forms,
2. test–retest, and/or the
3. split-half techniques.

If Form A and Form B are found to be equivalent when administered to the same group the test is reliable. Administering a test, waiting some time, and then administering the same test again should produce approximately the same results. Some adjustment may need to be made for the learning which has taken place between test administrations. The split-half technique is similar to the two equivalent forms approach for determining reliability and is arrived at through computation using a statistical formula.

A simple way to increase the reliability of the test is to increase the length of the test. Reliability is affected by the time the test is taken, how the individuals taking the test feel, and other environmental factors. It should be remembered that a test can be reliable without being valid.

Discrimination. A good test identifies the students who do well from those who perform less well. If the test does not discriminate in this manner it is not performing the function for which it was designed. If the most capable students in the subject do best on the test then the test is discriminating positively. If the least capable students do best on the test the test is discriminating negatively.

Objectivity. If the students understand and properly interpret the items of the test, the test may be said to be objective. The test is then a fair test for the sudents. Objectivity also applies to the scoring of the test papers by the instructor. In this sense the consistent scoring of the items of the test is the factor being considered. It is much easier to be objective in scoring a multiple choice question than an essay question.

Comprehensiveness. Enough items must be included in the test to properly sample from each of the units of the course. Inclusion of a larger number of items increases the comprehensiveness of the test. Care must be taken to include all units or major portions of the course. The task is greatly simplified if performance goals are developed at the operations or task level and test items built to include the identified performance goals.

A table of specifications may be developed as an aid in making a more effective plan of sampling of items. Usually all of the performance goals are not specifically included in the test items used.

Usability. Is the test easy to use; that is, to administer and score? These features increase the desirability of the test from the point of view of the instructor.

WHAT AND HOW OF TESTING

Refer to the performance goals designed for the lesson, unit, and course. These performance goals should indicate what the student is doing when he is performing or doing that which he is expected to do when demonstrating that he has acquired the skills, knowledges, or habits identified. Then, the performance goals must result in establishing a criterion of successful performance. This is usually in terms of quantity and quality.

The evaluation of performance is then a matter of asking, "What am I trying to measure?" This should be clearly evident from the performance goals. The next question is, "How am I going to measure it most effectively?" This must be determined in light of the kind of available device for securing the information. In some cases the decision might be to use an objective type of test item; whereas, in other cases an essay question might be more appropriate. In still other cases, a performance test may be given; while in other cases, assessment may be most valid in terms of direct observation of the students at work.

CONCEPTS IN TEST CONSTRUCTION

Some of the basic concepts in test construction which will improve the performance of evaluation are:

Determine best type of test for purpose. No one kind of test or method of evaluation of performance is best for all situations. Select that type which best provides valid, reliable, comprehensive, and objective information in harmony with the performance goals for which appraisal is sought.

Avoid giving away the answer. Select items that are independent of the other items. Include no items for which the answer is obvious to a person who does not know the subject matter. Be careful not to give clues to the students through the choice of words used. When developing items for a multiple choice test, use a minimum of four distractors.

State the question or item clearly. Keep the wording simple and concise. Eliminate the use of ambiguous items. Avoid the use of trick questions in the test.

Emphasize important items. Select items or questions that are important for the student to know or be able to do. Avoid insignificant details, or minor points dependent mainly upon memory. Use situation-type items, and items requiring problem-solving ability similar to that required on the job.

Avoid using negative items. Place the emphasis on the positive. Put the focus on what to do, not on what *not* to do. The reward must be given for correct action rather than for incorrect action!

Test bank. Build a bank of test items which can readily be used in assembling another form of the required test. Use of three-by-five-inch cards with the performance goal on one side and the test item with the correct response on the other is highly desirable.

Integration of Evaluation into a System of Instruction

No system of instruction is complete without some form of evaluation. Evaluation of performance is vital to improvement of the instructional system as well as essential for determination of the success of the student. Evaluation of performance must grow out of the identification of the performance goals and the content of the curriculum designed to supply the skill, knowledges, habits, and attitudes essential for entry-level employment in vocational and technical fields. Effective performance evaluation must reflect that which the student will need to know and do on the job. Therefore, application is the key word to determination of competencies for employment and fulfillment of the terminal objectives of the instructional program.

Summary

Evaluation of students' performance must be in terms of the performance goals established.

Evaluation may be in terms of:

1. Process evaluation—success at the time of progression through the educational program.
2. Product evaluation—success on the job after completion of the program of instruction.

Evaluation has many purposes. Some of the main purposes are:

1. Determination of qualifications for entry-level positions.
2. Improvement of the instructional program.
3. Awarding of grades for promotional and administrative uses.

Evaluation may employ an absolute criterion or a relative base of performance. The absolute criterion is usually geared to the demands of entry-level jobs.

Many sources of information provide the foundation for evaluation. These include paper-and-pencil tests, performance tests, and evaluation of the completed project. Each method must be selected for the needs to be served, and the suitability of the method for the situation.

A good test is:

1. Valid—measures what it is supposed to
2. Reliable—consistent
3. Discriminatory—separates degrees of accomplishment
4. Objective—fair to student
5. Comprehensive—samples from all units taught
6. Usable—easy to give and score

Careful selection or development of an appropriate instrument is the key to evaluation. The evaluations must be directly related to the performance goals established.

QUESTIONS AND ACTIVITIES

1. What are the purposes of evaluation? Discuss.
2. Explain the differences between evaluation, measurement, and assessment.
3. What is the difference between process and product evaluation? Discuss the merits of each.
4. What is meant by an absolute criterion for evaluation? Explain.
5. Which is more appropriate in vocational and technical education, the absolute criterion or the relative-based method? Why?
6. Is a paper-and-pencil test an effective way of evaluating the progress of the students in a shop class? Discuss. How could this be done more effectively?
7. What are the purposes of: achievement tests, aptitude tests, interest tests?
8. List and explain the characteristics of a good test.
9. Can a test be reliable without being valid?
10. Should a test provide information as to discrimination?
11. What are the two most fundamental questions to ask in planning a test?
12. What principles should be considered in constructing a test? Discuss each.
13. How is performance evaluation related to the instructional process? To the performance goals?
14. Diagram (as a flow chart) a systems approach to vocational and/or technical education. Indicate on the diagram, performance goals, curriculum elements, and performance evaluation. Make this a closed-loop system with performance evaluation providing the "feedback."

REFERENCES

1. Lessinger, Leon M., and Dwight H. Allen, "Performance Proposals for Educational Funding: A New Approach to Federal Resource Allocation." *Phi Delta Kappan* 51 (November, 1969).
2. Micheels, William J., and M. Ray Carnes, *Measuring Educational Achievement.* New York: McGraw-Hill Book Company, Inc., 1950. See pp. 104–23.

Chapter 16 DEVELOPING AND USING OBJECTIVE-TYPE TESTS

Evaluation of performance of the student and improvement of the teaching–learning process are two basic reasons for giving tests. Tests may be paper-and-pencil tests or application-type tests administered in the shop or laboratory.

Paper-and-pencil tests are described as objective tests or essay tests. Objective tests are of two kinds: recognition or recall.

Types of Objective Test Items

Recognition test items are multiple choice, true–false, or matching items. The student recognizes the correct response, or fails to recognize the correct response, from two or more possibilities that are given as part of the question. Some recall is essential in the recognition of the correct response; however, the scope of the student's ability to recognize is much broader than recall ability.

Objective recall tests consist of completion, listing, or some modification of these forms. The student is required to supply the information requested rather than just identify it as required in the recognition type of test items.

In each case a good test item must be valid, reliable, and objective. The test must be comprehensive and needs to discriminate if it is to serve well the functions for which it is designed. Application rather than verbalization is necessary if the test item is to evaluate the performance capability of the student.

CRITERIA FOR EFFECTIVE TEST ITEMS

Each test item has a vital role to play in the job of ascertaining the quantity and quality of student performance on the subject matter being tested. To serve this function well, the test must be directly connected with a performance goal (or behavioral objective) written at the level of the operation or task and "geared" to identified needs growing out of the activities actually performed on the job. In fact, one method is to develop one or more test items as a companion item for each performance goal at the time the performance goal is being determined.

CHECK QUESTIONS FOR ITEM CONSTRUCTION

In building test items it is helpful to ask the following questions:

1. Does this item measure the student's achievement of a specific performance goal established as fundamental to employment on the job? (Have one or more items for each goal.)
2. Does this test item use application of knowledge rather than just memorization of textbook information?
3. Is the item clearly written in language understandable to the student?
4. Does the item test an important rather than a trivial element?
5. Is the test item realistic in accordance with the requirements of the job for which the student is being prepared?
6. Is the test item well planned and carefully written?
7. Are the English and the format correct?
8. Does the content of the test items parallel the work of the class?
9. Does the test item discriminate between students who have mastered the content in varying degrees? (Even though an absolute criterion is used variations in the degree of accomplishment will be evident.)

Test Bank or Resource File

A collection of test items into a test bank will serve as an excellent resource when a new test is to be developed or an existing test is to be revised. These items should be started at the time the performance goals are being written. Additional items are easily added as the course content is identified. Each time the course is repeated new items will come to mind for the test bank. Likewise, some items will be withdrawn as ineffective after making the item analysis following the giving of the test to the students.

A "direction" card needs to be written for each kind of item. Illustrations of the content of such cards follow:

Multiple choice direction card

Each of the following or incomplete statements given below is followed by several possible answers. Choose the part that *best* answers or completes the statement. Circle the identifying letter only of the correct answer on the answer sheet opposite the corresponding number.

True–false direction card

If the statement is true, circle the T on the answer sheet; but if it is false, circle the letter F. Use corresponding numbers.

CARDS FOR THE TEST BANK

The basic format for the cards for the test bank is illustrated in Exhibit 16–1. Modifications of the system may be made in keeping with the desired objectives.

ANSWER SHEET FOR HAND-SCORED OBJECTIVE TESTS

Exhibit 16–2 suggests an answer sheet format that can be used with a variety of objective testing. The form provides space for responding to multiple choice items in Part I, and true–false items in Part II. Part III can be used for either completion or matching. The back side of the answer sheet could be numbered for brief essay responses.

To expedite scoring, prepare a correct answer sheet. A transparency may be made of this answer sheet and laid over the answer sheet containing the responses of the student.

It is also possible to make a simple mechanical scoring machine for multiple choice and true–false items using a wooden form and metal pegs positioned for the correct responses to each item.

Another common aid consists of a stencil designed to permit only the correct answer to show. This can be made by correctly marking a standard answer sheet and then punching out the marked answer with a leather punch or other suitable tool.

Multiple Choice Items

Evaluation of knowledge essential to determination of the performance capability of the student may be achieved by the use of an objective test consisting of multiple choice items used in conjunction with other forms of objective or essay items.

EXHIBIT 16–1

Test Question Card System

1. Develop test questions and place one question on each card (3″ x 5″).
2. Identify all question cards by coding in accordance with the following or another desired system.

FRONT OF CARD

Course No. and Name	Unit and Job (or lesson) (Book, author, page)
QUESTION (Centered)	
	Correct response

BACK OF CARD

Your Name	Date
PERFORMANCE GOAL SERVED BY QUESTION	
	Assignment number and name

EXHIBIT 16–2

ANSWER SHEET

NAME__________________________________DATE______________

COURSE NUMBER AND NAME __________________SCORE__________

PART I			PART II			PART III
1. a b c d	26. a b c d	51. a b c d	76. T F	101. T F	126. T F	151._____
2. a b c d	27. a b c d	52. a b c d	77. T F	102. T F	127. T F	152._____
3. a b c d	28. a b c d	53. a b c d	78. T F	103. T F	128. T F	153._____
4. a b c d	29. a b c d	54. a b c d	79. T F	104. T F	129. T F	154._____
5. a b c d	30. a b c d	55. a b c d	80. T F	105. T F	130. T F	155._____
6. a b c d	31. a b c d	56. a b c d	81. T F	106. T F	131. T F	156._____
7. a b c d	32. a b c d	57. a b c d	82. T F	107. T F	132. T F	157._____
8. a b c d	33. a b c d	58. a b c d	83. T F	108. T F	133. T F	158._____
9. a b c d	34. a b c d	59. a b c d	84. T F	109. T F	134. T F	159._____
10. a b c d	35. a b c d	60. a b c d	85. T F	110. T F	135. T F	160._____
11. a b c d	36. a b c d	61. a b c d	86. T F	111. T F	136. T F	161._____
12. a b c d	37. a b c d	62. a b c d	87. T F	112. T F	137. T F	162._____
13. a b c d	38. a b c d	63. a b c d	88. T F	113. T F	138. T F	163._____
14. a b c d	39. a b c d	64. a b c d	89. T F	114. T F	139. T F	164._____
15. a b c d	40. a b c d	65. a b c d	90. T F	115. T F	140. T F	165._____
16. a b c d	41. a b c d	66. a b c d	91. T F	116. T F	141. T F	166._____
17. a b c d	42. a b c d	67. a b c d	92. T F	117. T F	142. T F	167._____
18. a b c d	43. a b c d	68. a b c d	93. T F	118. T F	143. T F	168._____
19. a b c d	44. a b c d	69. a b c d	94. T F	119. T F	144. T F	169._____
20. a b c d	45. a b c d	70. a b c d	95. T F	120. T F	145. T F	170._____
21. a b c d	46. a b c d	71. a b c d	96. T F	121. T F	146. T F	171._____
22. a b c d	47. a b c d	72. a b c d	97. T F	122. T F	147. T F	172._____
23. a b c d	48. a b c d	73. a b c d	98. T F	123. T F	148. T F	173._____
24. a b c d	49. a b c d	74. a b c d	99. T F	124. T F	149. T F	174._____
25. a b c d	50. a b c d	75. a b c d	100. T F	125. T F	150. T F	175._____

The multiple choice item is frequently used since it can be effectively developed to measure the student's ability to interpret, discriminate, select, and make application of content learned. This item also can be used to measure judgment, understanding, and inferential reasoning ability. The ability of the student to recognize a correct response can be determined by this type of test item. One of the advantages of using this type of item is the fact that the scoring can be made entirely objective.

The multiple choice item is frequently used for standardized tests as well as for teacher-made tests. Multiple choice items are easily machine scored.

FORMAT

The multiple choice item consists of a stem which contains the central problem or question, which is followed by a series of possible answers. Possible incorrect answers given are referred to as distractors or decoys. Four or more possible answers should be provided for each stem. Each of the distractors should be a possible response for the student who is not sure of the correct answer. The statement given in the stem should be a clear, concise, and specific expression of the central problem. The organization of the item should be as follows:

_____1. A screw is an illustration of the principle of the
a. inclined plane
b. wheel and axle
c. lever
d. wedge

_____2. In choosing a thermocouple for use in measuring rapidly changing temperature, which of the following factors is most important?
a. linear output
b. resistance to corrosion
c. repeatability
d. size of wire

One right answer type of multiple choice item. The student is given a stem and several possible choices of which only one is correct. He is required to select the correct response.

Best answer type of multiple choice item. This type of multiple choice item requires the exercise of judgment on the part of the student. Each of the choices given may be partially correct. The student must demonstrate his ability to choose the best response as related to the central problem given in the stem. This is considered the most effective form of the multiple choice items.

Association items. This is used to measure the inferential reasoning ability of students. It is frequently used in standardized tests. The student must select the word, phrase, or illustration from the possible choices which is most closely associated with the part given in the stem.

Analogy-type multiple choice item. This type of item uses the same format and concept of a mathematical proportion. The relationship existing between the first two parts of the item must be determined by the student and translated into the relationship between the last two parts. The third part is given in the stem while the fourth is to be selected from among the choices provided.

Reverse multiple choice item. In this type of item the student is required to choose the poorest or the wrong response from among several correct responses. Reverse multiple choice needs to be used with discretion. It is a negative rather than a positive approach to the problem. Good psychology and good teaching suggest that in the main the positive approach should be used. There are, however, instances in which it is easier to construct an item with three or four responses that are true rather than to make several distractors false. If such an item is deemed desirable and the word "not" or a similar word is selected, it is important to underline the word or write it in full capital letters so the student will be sure to note the fact that this is a reverse multiple choice item.

ILLUSTRATIONS OF MULTIPLE CHOICE ITEMS

The following multiple choice items illustrate some of the types frequently used in teacher-made tests:

AUTO MECHANICS

1. Cylinder bore taper is determined by measuring differences from
   A. top of ring ridge to bottom of cylinder
   B. just below ring ridge to bottom of cylinder
   C. front to rear and side to side of bore
   D. below ring ridge to bottom of ring travel

(Correct response is *B*)

ELECTRONICS

2. 100-ohm, 200-ohm, and 300-ohm resistors are connected in series across a 100-volt supply. The voltage across the 200-ohm resistor will be
   A. 22.5 V
   B. 33.3 V
   C. 50 V
   D. 67 V

(Correct response is *B*)

3. Three capacitors each having a value of 10 μF are connected in parallel. The total capacitance is therefore
   A. 5 μF
   B. 10 μF
   C. 20 μF
   D. 30 μF

(Correct response is *D*)

4. The current required for full-scale deflection on a 20,000-ohm voltmeter movement is
   A. 50 μA
   B. 500 μA
   C. 1.00 mA
   D. 5.00 mA

(Correct response is *B*)

5. An attenuator measures 8.1 dB and has 3 mW at its input. What is the power at the output?
   A. 46.5 mW
   B. 4.65 mW
   C. .465 mW
   D. .0465 mW

(Correct response is *C*)

6. What is the maximum current which can be safely passed through the entire length of a linear adjustable 25-ohm, 50-watt resistor, 4 inches long?
   A. 0.707 A
   B. 1.414 A
   C. 2.000 A
   D. 2.828 A

(Correct response is *B*)

PRACTICAL NURSING

7. Mrs. Brown is weak from lack of food after ironing all morning. To obtain quick energy she should take
   A. a doughnut
   B. some prunes
   C. a ham sandwich
   D. a spoon of peanut butter
   E. some orange juice

(Correct response is *E*)

8. Sue is 5 feet, 7 inches tall, of a small frame, and weighs 56 kilograms. This means that her ideal weight is
   A. 109.4 lb
   B. 112.5 lb
   C. 123.2 lb
   D. 132.2 lb

(Correct response is *C*)

CARPENTRY

9. You are to lay out the stringers for the stairs. The total rise is 8′4″ and there are 13 treads with a unit run of 10⅜″. What is the *unit rise* to the nearest ⅛″?
   A. 7¾″
   B. 7⅝″
   C. 7⅜″
   D. 7⅛″

(Correct response is *D*)

10. You are preparing exposed rafter tails for painting. When knot holes are encountered you should
    A. fill with putty
    B. fill with texture
    C. fill with plastic wood
    D. fill with spackling

(Correct response is *D*)

11. What is the length of the header for a 3′0″ door using a ¾″ jamb?
    A. 3′4¼″
    B. 3′5⅜″
    C. 3′6⅛″
    D. 3′7½″

(Correct response is *B*)

12. If a roof has a rise of ten feet and the run is twelve feet, what is the rise per foot run?
    A. 6″
    B. 8″
    C. 10″
    D. 12″

(Correct response is *C*)

HEAVY EQUIPMENT OPERATING AND MAINTENANCE

13. A hydraulic coupling has an oil chamber containing a web of pump vanes driven by the engine. Connected to the driven machinery is a set of
    A. turbines
    B. rotors
    C. stators
    D. impellers

(Correct response is *A*)

True–False Items

The true–false item consists of a sentence which is to be identified by the student as right or wrong. A situation from the job environment can be built into such an item. Opportunity for the student to exercise judgment is one of the characteristics of a well-written true–false item. Since there are only two possible choices in the regular true–false item it provides an opportunity for the student who does not have the correct response to guess and receive approximately one-half of the items correct.

Regular true–false item. This is the most common form of true–false item. The student reads the sentence and responds by marking either "true" or "false" or the appropriately designated symbol.

Cluster true–false item. The cluster type consists of an incomplete statement followed by several parts, each so phrased as to complete the stem. The student marks those options which correctly complete the stem. The item may be so designed that one or more parts is correct for each item.

Modified true–false item. This type of true–false item reduces the opportunity for the student to guess. This item is usually designed so that the student must do two things. First, he must decide whether or not the item is true. Then, if the item is false, he must indicate the part that is false and change that part to make the statement true.

FACTORS IN THE CONSTRUCTION OF TRUE–FALSE ITEMS

It is frequently difficult to construct true–false items that are entirely false. Students are frequently led to guess on true–false items. Many times true–false items are based on unimportant details of questionable significance. In building true–false items:

1. Have half or more of the items true.
2. Use items that test applications of knowledge.
3. Express sentences in clear, concise, and specific language.
4. Avoid giving clues. Use of words such as "always" or "never," or having true statements longer than false statements provides such clues.
5. Use modified true–false items rather than plain items whenever possible.

ILLUSTRATIONS OF TRUE–FALSE ITEMS

Study of the following true–false items should be helpful.

WELDING

T F 1. When welding mild steel with a neutral flame, one of the visible evidences of complete penetration is a large volume of sparks issuing from the molten puddle.

(The correct response is *false*)

AUTO MECHANICS

T F 2. A lean mixture of fuel and air will cause the engine to run cool.

(The correct response is *false*)

PRACTICAL NURSING

T F 3. Preparing the formula using clean bottles and equipment and then sterilizing in covered bottles is known as the aseptic method of sterilization.

(The correct response is *false*)

CHEF'S TRAINING

T F 4. Hands provide one of the most common methods for the transfer of bacteria from one place to another.

(The correct response is *true*)

T F 5. Bread may be kept fresh by placing in the refrigerator.

(The correct response is *false*)

TELEVISION SERVICING

T F 6. Additive hum that produces two hum bars on the picture is produced by improper filtering in a full-wave rectifier power supply.

(The correct response is *true*)

T F 7. If the vertical hold control can stop the picture from rolling only temporarily the trouble is probably the oscillator.

(The correct response is *false*)

APPLIANCE REPAIR

T F 8. A circuit with an applied voltage of .5 kV and a measured current of 100 mA has a resistance of 5 K.

(The correct response is *true*)

T F 9. When mixing acid and water, pour the acid into the water.

(The correct response is *true*)

DESCRIPTIVE GEOMETRY

T F 10. A line parallel to a plane shows parallel to an object line of the plane in all views.

(The correct response is *false*)

T F 11. An auxiliary view is projected perpendicular to a profile reference plane.

(The correct response is *false*)

Matching Items

Matching items are a variation of multiple choice items. Instead of selecting one choice among four or five possibilities the student has two columns or groups to match. Matching is effective for determining the student's ability to recognize relationships and make associations. Since several parts are assembled in each group guessing can be greatly reduced.

USES

Matching is used to test the student's knowledge in situations such as the following:

1. Definition of terms
2. Relationship of parts
3. Classifications of objects
4. Cause and effect elements
5. Symbols and meanings of symbols
6. Nomenclature and illustrations of parts

THINGS TO REMEMBER IN BUILDING MATCHING ITEMS

Better matching items will result if you remember to:

1. Provide specific directions for the student.
2. Have the longer item on the left in a numerical sequence with the shorter item on the right using a lettered designation.
3. Include several extra choices from which correct choices must be made.
4. Keep the entire item, both right- and left-hand columns, on one page.
5. Use related concepts in one exercise.
6. Have at least six but less than twelve items in each exercise.

ILLUSTRATIONS OF MATCHING ITEMS

The use of matching items in tests of vocational education is illustrated by the items which follow relative to carpentry, oxyacetylene welding, and practical nursing.

CARPENTRY

DIRECTIONS: Place in the blank space at the left the letter which identifies the correct choice of the words necessary to match the given phrase. The first is completed as a sample.

Column I	Column II
E x. Interior walls	A. Firestop
___ 1. Support used when joists are nailed to studs	B. Sheathing
___ 2. Solid bridging in stud wall	C. Header
___ 3. Material nailed to exterior walls for rigidity	D. Ribbon
___ 4. Header in the exterior wall rests on this 2″ × 4″	E. Partitions
___ 5. 2″ × 4″ supporting studs	F. Trimmer
	G. Sole plate
	H. Top plate

(Correct responses: 1–D; 2–A; 3–B; 4–F; 5–G)

OXYACETYLENE WELDING

DIRECTIONS: Match the part in Column II with that in Column I. While some of the items in Column II may not need to be used, others may be used more than once.

Column I

___ 1. Is a material that will permanently deform without failure.
___ 2. Is a material that can be rolled into thin sheets.
___ 3. Is where the test specimen load is increased slowly; a point is found at which a definite increase in length results with no increase in load.
___ 4. Is the pounds per square inch required to produce a certain amount of permanent elongation.
___ 5. Is a material that can withstand sudden or often applied forces and will deform before failure.
___ 6. Shows no permanent distortion before failure.
___ 7. Is a material resisting penetration.

Column II

A. Elasticity
B. Elongation
C. Toughness
D. Contraction
E. Tensile strength
F. Malleability
G. Ductility
H. Expansion
I. Brittleness
J. Yield strength
K. Hardness
L. Yield point
M. Rupture point

(Correct responses: 1–G; 2–F; 3–L; 4–J; 5–C; 6–I; 7–K)

PRACTICAL NURSING

DIRECTIONS: Place on the blank space at the left the letter which identifies the correct choice of the words necessary to match the numbered statement. Use each letter only once. A sample is given below.

Column I

D x. The acute gastric ulcer
___ 1. The acute gall bladder
___ 2. The new postoperative gastrectomy
___ 3. The spastic colon
___ 4. The second-degree burn
___ 5. The edematous cardiac
___ 6. The general postoperative diet
___ 7. The gout
___ 8. The atonic constipation
___ 9. The diabetic
___10. The geriatric without dentures

Column II

A. High salt diet
B. Low salt diet
C. Low residue diet
D. Sippy diet
E. Low fat diet
F. Dry diet
G. A.D.A. diet
H. High purine diet
I. Low protein diet
J. Low purine diet
K. High residue diet
L. Clear liquid diet
M. Mechanical soft diet

(Correct responses: 1–E; 2–F; 3–C; 4–A; 5–B; 6–L; 7–J; 8–K; 9–G; 10–M)

Completion Items

The student is required to fill in the missing blank or blanks in this type of item. These items are frequently described as simple recall or as free-response items. Several variations are possible using the same general framework.

Completion items are considered good for measuring retention of specific points. It is definitely recommended that if recall items of this type are to be used the emphasis should be placed on applications of knowledge.

Completion items are easy to construct but due to the nature of the item are more difficult to score. The student may interpret incorrectly the item or may find it difficult to select the word which was intended for the item. This means that scoring, if it is to be fair, must be rather subjective.

FACTORS TO CONSIDER

In building completion items remember to:

1. Use clear, positive statements.
2. Place the blanks near the end of the statement.
3. Construct the item so there is only one correct response.
4. Avoid the inclusion of clues, i.e., "A," "an," etc.

USE

The use of completion items should be limited to applications where there is no question as to the meaning intended and where the nature of the proper response to be placed on the blank is clearly evident, if the student knows the content on which he is being tested. This type of objective test item is often the most frequently misinterpreted and misunderstood by the student. This is especially true if several blanks are to be completed in one sentence. Extreme care needs to be exercised in the construction of completion items.

ILLUSTRATIONS

Study the following completion items:

AUTO MECHANIC

DIRECTIONS: Write the correct answer on the blank.

1. An intake valve on an F head engine is located in the ________. (head)
2. We can compensate for the metal we grind off a valve and seat by installing ________. (shim)

3. If the rotor in the distributor will turn a little clockwise, the engine direction of rotation will be __________. (clockwise)

CARPENTRY

DIRECTIONS: Write the correct answer on the blank.

1. The most common spacing of rafters is __________. (24″ on center)
2. The small roof structure that sheds moisture from behind chimneys is called __________. (a saddle)
3. The type of roof that is made of two flat surfaces intersecting at a ridge is called __________ roof. (a gable)

DESCRIPTIVE GEOMETRY

DIRECTIONS: Write the correct answer on the blank.

1. A line parallel to a plane will show parallel to __________. (a line on the plane)
2. The true angle between two planes will be measured when both planes __________. (are on edge)
3. Parallel lines will show parallel to each other in __________. (all views)

NUTRITION

DIRECTIONS: Write the correct answer on the blank.

1. Liver, egg yolks, lean meat, and molasses are good sources of the mineral __________. (iron)
2. 154 pounds of body weight are equal to __________ kilograms. (70)
3. Polyunsaturated fats are mainly available from __________. (vegetable oils)

WELDING

DIRECTIONS: Write the correct answer on the blank.

1. A chemical liquid in acetylene cylinders that absorbs acetylene is __________. (acetone)
2. A safety precaution for welding a container that has had flammable material in it is to fill the container with __________. (water)
3. Braze-welded parts lose their strength and ductility when subjected to temperatures over __________ degrees F. (500)

ELECTRONICS TECHNOLOGY

DIRECTIONS: Write the correct answer on the blank.

1. The image frequency of a 600 kc signal is __________. (1510 kc)
2. The oscillator in a superheterodyne receiver is oscillating at 1255 kc. The station frequency of the receiver signal is __________. (800 kc)
3. The characteristic of the logarithm of 3.1416 is __________. (0)

Summary

Tests are useful for evaluation of the student's ability to perform. They also provide feedback for improving the teaching–learning process.

Objective-type tests are either recognition or recall. Multiple choice, true–false, and matching are recognition-type tests. Completion and listing are recall-type tests.

Each test item should be derived from a previously stated performance goal. The test item should measure the student's ability to perform rather than just to memorize.

A test bank of many items, each written on separate three-by-five-inch cards, provides an excellent resource file for building or revising tests. A direction card should be developed for each different type of item.

Multiple choice items are effective in measuring the student's ability to interpret, discriminate, select, and apply knowledge. The best answer type of multiple choice item is considered most effective as it necessitates the use of judgment on the part of the student.

The modified true–false item reduces the possibility of guessing. In building true–false items test for application of knowledge; be specific. Avoid giving clues. Have more true than false items.

Matching items are designed to determine the student's ability to recognize relationships and make associations. This type is effective for testing definitions, classifications, symbols, nomenclature, and relationships.

While completion items are satisfactory for measuring retention, the student may have trouble deciding what word(s) to use. Design items so that only one response correctly completes the sentence. Use situation and application items whenever possible in objective tests.

QUESTIONS AND ACTIVITIES

1. What is an objective test?
2. Contrast the recognition- and the recall-type of objective tests.
3. Explain the advantages and the disadvantages of the following types of test items:
    - A. Multiple choice
    - B. True–false
    - C. Matching
    - D. Completion

4. Explain the relationship of the performance goal to the test item in a well-designed test.
5. What questions may help the test item writer to check the quality of his work?
6. Why is the test bank a valuable tool?
7. What is the purpose of the:
   A. Direction card?
   B. Answer sheet?
8. Identify and describe the features of the different types of:
   A. Multiple choice test items
   B. True–false test items
9. What types of situations are suited to matching items?
10. What difficulties are frequently encountered in responding to completion-type test items? How can this be avoided?
11. Develop the following application-type test items, harmonizing each item with a performance goal for one of the courses you are teaching or may teach in the future.
    A. Multiple choice—50 items
    B. True–false—30 items
    C. Matching—2 sets of items
    D. Completion—10 items
12. Write appropriate direction cards for each different type of test item mentioned in question 11.
13. Assemble test items developed in question 11 into an appropriate test for the students in the course selected. Design for use with a standard or individually designed answer sheet. Prepare the key for the test.

Chapter 17 WRITING ITEMS FOR SHORT ANSWER AND ESSAY TESTS

Essay and short answer tests are frequently described as subjective in contrast to the multiple choice, true–false, and matching types which are classified as objective tests.

Essay tests are frequently used by teachers who prepare brief tests for use in individual classes. The objective tests have also become widely used as a teacher-made test as well as for standardized tests. Significant functions of evaluation can be served by essay tests.

Advantages of the Essay Test

The assessment of the advantages of the essay test must be directly related to the performance goals identified. The performance goals of the test items are or should be closely correlated with the performance goals identified for the units of the course. The objectives of the performance goals of the course in turn should be an outgrowth of the total objectives of the educational program.

An essay test has considerable merit if the questions are carefully planned and written. Frequently at the graduate level in colleges and universities essay questions are extensively used.

The strengths of the essay questions as instruments for evaluation of information comprehension and application of knowledge are established by the following introductions to performance goals:

To organize—
To defend—
To apply the scientific method to the solution of a problem—
To explain the relationships—
To contrast the strengths of the following systems—
To compare—
To weigh the relative merits of the following—
To arrive at the best decision considering the available solutions and defend your choice—
To describe—
To tell how—
To trace the flow of—

Much of the criticism which has been directed against the essay test should instead have been directed at the quality of the test items and how these have been ineffectively used. While many essay questions may not have been well designed to utilize the strengths and advantages of such a question, the same criticism can justly be made of items on objective tests.

Effective essay questions need to be designed to evaluate the ability of the student to:

1. Express himself effectively.
2. Organize the elements of the response into a comprehensive and effective answer.
3. Describe, explain, compare, or contrast in meaningful terms to demonstrate applications of knowledge desired by the author of the test question.

Essay questions lend themselves to evaluation of problem solving and situation experiences. This reflects a higher type of mental ability than simple recall. Recall is part of the process, but reasoning and selecting of facts and focusing these facts upon the issues and concerns present much greater challenges than occur when the individual is merely required to select a "canned" response from one or more possible answers already given to him. The student is confronted with the need to reason, plan, organize, and express his responses to the question.

Disadvantages of the Essay Test

The main criticism against the essay test is focused on the difficulty of objective scoring or evaluation. Students often write much irrelevant material which has little direct bearing upon the question. How such a response is to be evaluated becomes a rather difficult question.

Various instructors will give different appraisals of the student's response. This results in a lack of objectivity. Such test items may then not reflect the true knowledge or ability of the student. This may not be fair to the student. The major difficulty is usually with the test items that have been used rather than with the type of item.

Formulating Valid Essay Questions

Valid essay questions can be constructed! First, the performance goals sought must be carefully delineated. Use essay questions only for such items as lend themselves to essay questions. Establish a meaningful performance goal (behavioral objective). Then phrase the question to focus specifically on that performance goal. Test the wording of the question to determine whether or not the meaning is clear. Reread the question to ascertain whether the statement is effectively delimited to the concepts desired. Check to see whether or not you have called for specific answers. Avoid broad, general, meaningless questions which lack "focus." Require the student to describe, explain, compare, or contrast. Insist on effective expression and organization of the response. Gauge the level of response which your students are capable of or should be capable of making, and pre-establish the basic key elements which should be covered in the response. Having thus identified the criteria of performance, a sliding scale can be utilized to recognize individual differences in the student's responses. Gradations of acceptable performance can be determined, as well as the level of performance which is not acceptable.

ASSESSING STUDENTS' RESPONSES

The test reader must prepare in advance for objectivity if objectivity is to be achieved in the evaluation of essay or short answer test items. After the questions have been critically assessed, the key elements of the acceptable response must be determined. In making this determination the relative emphasis on application, memorization, and other possible response categories and elements should be made.

After establishing a master response for each question, provision can and should be made for deviations. A tolerance of plus and minus can be adopted much the same as that used on blueprints for machine parts to be manufactured in industry. The tolerance must suit the identified performance goal(s) for the question.

The directions provided for the test questions must be clear, specific, and definite so that the students understand what they are expected to do.

In rating the questions the test reader must be as objective as possible. If a four-question essay test, completed by twenty-five students, is to be evaluated objectively, the test reader should evaluate all of the number one questions at one time. The rating should be done in accordance with the pre-established specifications for marking. After reviewing the twenty-fifth paper it is desirable to review again the first few papers to identify any variance in values given to the student's questions. While the above procedure will usually improve objectivity, there is much possibility of variation unless a great deal of care is taken by the test reader. The use of several test readers for the same papers has a tendency to increase objectivity. It would then be necessary to reconcile the ratings between the scores.

ILLUSTRATIONS OF ESSAY QUESTIONS

The following questions are illustrations of questions suitable for use in vocational and technical classes. These questions were developed by students in courses in performance evaluation:

1. Explain why fluxes are used in the brazing process.
2. Describe the procedure to follow in keeping drawings clean in the drafting room.
3. Plan a menu for breakfast using the "basic four" food pattern. Include standard food portions for servings for six people.
4. Describe the nature and the function of the two types of service valves used on automotive air conditioners.
5. Describe the fetal circulatory system as it exists in the month prior to birth.
6. The latest financial statement of one of your customers provides the following information:

   Cash, $356,500; receivables, $920,500; merchandise, $559,500; current debts, $475,600; deferred, $89,700; net worth, $2,483,-400; sales, $3,358,000. The operations have been profitable during the last two years, earning $82,000 for the last year and $102,600 in the previous year. Three years ago a loss of $53,400 was incurred, and four years ago the loss was $41,700. Annual dividends of $18,800 have been paid for the past five years. Sales volume four years ago reached an all-time peak. Shortages of merchandise brought about recessions in subsequent years, but this trend has been checked. Inventory turnover is active. Receivables are in good proportion to volume.

   Compute a credit limit for this firm. Explain in detail the basis you have used for arriving at your decision.

7. You are in the process of constructing a building. How would you determine that the corners are laid out square? Explain.
8. Compare the advantages of thread rolling to machining with a single-point tool.
9. Contrast the properties of high-speed steel tool bits with those of cemented-carbide tool bits.
10. You have selected a teflon plastic material as part of the design for a bushing for a slow moving shaft carrying a light load. Defend your choice.

Summary

Essay and short answer test items have been used successfully for evaluation of comprehension of information and application of knowledge.

The essay tests provide opportunities for students to:

1. Demonstrate ability of self-expression.
2. Organize effectively.
3. Treat the question according to the specification given, i.e., describe, explain, compare, or contrast.

The performance goal(s) must be carefully established and be consistent with that of the units of the course.

Essay questions often lack objectivity and are difficult to score. In writing essay questions:

1. Establish meaningful performance goals.
2. Focus the question relative to the established goal(s).
3. Provide a rigid specification so the student knows precisely what he is to do.
4. Require the student to describe, explain, compare, or contrast.
5. Establish the key elements for the correct responses prior to scoring the questions.
6. Plan criteria for and levels of quality that are acceptable.

QUESTIONS AND ACTIVITIES

1. Explain the characteristics of essay tests.
2. Compare the advantages of essay tests with those of objective tests.
3. Describe the essential characteristics of well-written essay test questions.
4. Identify performance goals suitable for essay test items.

5. How can the objectivity of essay questions be maximized?
6. Select several performance goals for units of one of your courses and build ten essay questions to evaluate achievement of these performance goals.
7. Establish the criteria for students' successful performance on the preceding questions. Gear the criteria to the grading system used, i.e., pass–fail, four levels of acceptable performance and one of failing performance, etc.

Chapter 18 EVALUATING THE FINISHED PRODUCT

In teaching related courses, the instructor may have less direct involvement than the shop or lab teacher with the shop or laboratory of the major. However, there are many instances in which direct involvement does occur. It is, therefore, essential that related instructors recognize the factors and problems essential in the evaluation of the finished product. How should the finished product be evaluated? How should the applications of related knowledges be determined? If the student cannot apply effectively those knowledges which he has acquired, has he learned successfully? What factors should be considered in making such an evaluation? These are important concerns worthy of further attention. This chapter identifies some of the problems and solutions that will be important to the teacher of related subjects as he helps his students understand the production process and standards of evaluating the finished product.

Identifying Standards

Both quality and quantity are significant elements of the output of the student. Employees in industry have criteria, in most cases, for determinations of success related to both these standards. The quality must be within acceptable tolerances as to such elements as size, finish, or other identifiable standards of excellence. Work which is not of suitable quality will be rejected, resulting in extra time and costs to correct or do over that which should have been done correctly in the first place. In some cases such errors result in the customer either rejecting the finished product or in returning it for additional corrective work. Both are expensive in terms of the unit cost of production and the reputation of the firm as well as the individuals producing the product or completing the task. Standards of quality are essential.

Industry and business can exist only if they can produce at a profit. A profit results only if the cost factors are maintained in proper perspective to the identified selling price. Therefore, predetermined levels of production must be maintained if a profit is to result. Techniques of cost accounting have resulted in the ability to forecast what these cost factors are and the levels of production essential to ensure the established ratio of selling price to cost. Some individuals working in industry achieve the desired level of quality but fail to produce adequate quantity to effect the desired profit for the firm. Students must learn early the importance of quality and quantity in relation to production.

Building Student Awareness of Required Standards

Effective teaching demands establishment of realistic goals. Goals are targets for students and teachers. But, even more, goals are destinations to be reached. From the very first day in shop or laboratory, students must be made aware of these targets. These targets provide built-in motivation and stimuli for action.

A sense of standards of job requirements results from the discussions and explanations of the instructor. Reference books frequently deal with specific tolerances used in industry. Some industries have manuals, such as flat-rate manuals, designed to communicate established time-rates for completion of the specific job.

Visits to industry create an awareness of the standards of production required for various types of activities. The emphasis varies. In some cases the major emphasis is on quality while in other cases a high rate of production with lesser but acceptable quality is demanded.

To build a concept of acceptable rate of production, it is usually recommended that in giving a demonstration the instructor proceed with the task at the normal rate first, then go through each step in slow motion so the student can better follow each of the steps of the process.

This method is designed to "pace" the job in the mind of the student. During the process references are often made to the quality standards required.

Service jobs as well as production tasks demand that the completed tasks be of acceptable quality. Completion of the tasks within predetermined time-frames results in maintenance of desirable levels of income. Failure to meet such standards results in loss of income and ultimately loss of the job for the employee.

ESTABLISHING A SYSTEM OF EVALUATION

The effective related subjects teacher establishes and maintains a system of realistic standards which aids the students in developing desired levels of performance.

Time is taken prior to student involvement to explain the standards for each task or job. As in industry, after the standards have been established and communicated to the student, evaluation follows to ascertain the degree of achievement. Achievement above or below accepted standards must be discussed and corrective measures taken, if the instructional program is to be effective.

BASIC ELEMENTS OF SYSTEM

Analysis of the jobs or tasks should result in clearly identifying the various elements of the criteria and the checkpoints to be used in making the determination of quality of the student's performance. Likewise, viewing the job from known rates of performance should provide both enabling and terminal goals for time standards. Once these have been determined, the next step is to communicate them to students in the group.

Participation of students in setting of standards has sometimes been used effectively when coupled with essential research of industries' requirements. Student involvement has a beneficial effect and often leads to high student morale.

Competition between individuals and groups within the shop or laboratory may be the spontaneous result of understanding the required standards. In other instances such competition grows out of the posting and maintaining of progress charts. Grading of completed jobs or projects also tends to stimulate competition among the students. Final grades influence the spirit of competition and give a businesslike atmosphere to the learning environment.

RECORDING PROGRESS

A record of the success of the student in achieving the established goals is essential. While the record may be kept by the instructor, there

EXHIBIT 18–1

PROGRESS CHART

SCHOOL____________________COURSE NO. AND NAME________

PERIOD____________________INSTRUCTOR____________

CODE: Job started Job finished Grade 2 5—highest 1—lowest passing JOB TITLES									
NAME									

are many advantages in having the student share responsibility for the record-keeping procedures.

A progress chart is necessary for keeping records. (See Exhibit 18–1.) The progress chart may be an individual chart maintained by each student and kept in his notebook. Many instructors prefer a system using a large progress chart on the wall. This chart may be used most effectively in conjunction with individual tickets for each job or task. The procedure may be as follows:

EXHIBIT 18–2

JOB ASSIGNMENT TICKET

JOB NUMBER________________________

JOB NAME__________________________

DATE______________________________

NAME______________________________

GRADE_____________________________

INSTRUCTOR________________________

1. Prior to beginning a new job (or task) the student completes a simple job ticket (see Exhibit 18–2) giving such information as:
   A. His name
   B. Date
   C. Title and number of job commenced

   He also indicates the beginning of the job on the progress chart.
2. When the job has been completed the student takes the job ticket to the instructor, who evaluates the performance of the student on that job. The instructor assigns the grade on that job in accordance with predetermined criteria and initials the job ticket.
3. The student then marks the progress chart to show completion of the job with the grade assigned and deposits the job ticket on the shop evaluation board adjacent to the progress chart for future reference.
4. The instructor periodically reviews the progress chart and the job tickets for each student. These become the original sources that will be used in determining the student's grade for shop or laboratory work completed.

Summary

Evaluation of the completed work of the student (job project) provides essential information for the total determination of the student's success. This process places the emphasis in a realistic perspective similar to that of industry.

Standards of achievement in the shop or laboratory are reflected in quality and quantity of successful accomplishment. The determination must be made against a known standard. Students must know the standards to be used for evaluation for maximum learning.

A system of recording needs to be built to reflect the achievements of the students. The progress chart and job assignment ticket are essential elements of such a system. These may be placed on a wall board prepared for the purpose.

Posted progress charts should be maintained by the students and reviewed at frequent intervals by the instructor.

QUESTIONS AND ACTIVITIES

1. How should shop or laboratory projects be evaluated?
2. Why should these jobs or projects be evaluated?
3. What standards or criteria should be established for the evaluation of the jobs or projects?
4. How does the laboratory or shop work provide a basis for self-motivation? How can the instructor maintain motivation at constructive levels?
5. Describe a system for recording and monitoring the student's progress in the shop or laboratory. Why is this plan desirable?
6. Develop the necessary elements of the system suggested in the preceding question for use in your own shop or laboratory class.
7. Should the students maintain the progress chart? Why?

Chapter 19 DETERMINING ABILITY TO PERFORM RELATED TASKS BY OBJECT AND/OR MANIPULATIVE-PERFORMANCE TESTS

A fundamental concept in vocational education expresses the thought that if the student cannot perform he has not learned. This is true in related subjects classes. This is true whether the student is reading a blueprint, making a sketch, solving a problem in mathematics, or building a prototype for a new model in the shop or laboratory. Many opportunities arise in related classes for involvement of the students in actual performance during the learning process. In such situations evaluation of learning may be accomplished more effectively by the use of object and/or performance tests.

Working with actual problems similar to those which the employer will assign to employees and using materials, equipment, machines, and/or other closely controlled conditions similar to those actually found in working on the job give a realism which is difficult to provide by a paper-and-pencil test. This method also provides a validation of the results achieved by the student which is much more dependable than that usually arrived at by written tests. For these and other similiar reasons object tests and manipulative-performance tests have much value in evaluating the ability of students in vocational and technical education to perform on the job and under conditions similar to those found on the job.

Object Tests

Object tests utilize actual objects as items of the test. The objects are displayed by the instructor and the student is asked to identify the object. He may also be required to indicate the use of the object and/or answer specific questions about the object.

All instructional programs using materials, supplies, or equipment can benefit by the use of the object test. Here are some sample applications for object test items:

Occupation	*Test Item*
Automobile mechanics	Identify the parts of a front wheel assembly.
Carpentry	Name the kinds of wood in the pieces of board displayed and tell whether open or close grain.
Machine shop	Name and give a use for each of the kinds of screw threads of the parts shown.
Machine shop	Using a vernier scale on the micrometer measure each part to the nearest one-thousandth of an inch and record your answer on your answer sheet in the space provided.
Drafting	Identify length of each part and translate into a scale of 1/8″ equals one foot.
Electronics	Identify each type of tube and indicate one purpose for which it was designed.

PREPARATION FOR ADMINISTRATION OF AN OBJECT TEST

The following steps suggest the procedure essential in preparing for an object test:

1. Determine the performance goals for which evaluation is desired.
2. Select the test items which will satisfy the performance goals.
3. Secure the parts or components essential for setting up the test station.
4. Establish test stations at convenient intervals allowing adequate space for students to work without interfering with other students.
5. Write the task statement for the student for each test station.
6. Indicate in the task statement the specifications as well as the task. Provide instructions for recording students' response.
7. Determine time to be allowed for the completion of each task by each student.

8. Establish standard of acceptable performance for the students on the object test.
9. Number each station.

ADMINISTRATION OF THE OBJECT TEST

Smooth administration of the object test is likely to result if the following are observed:

1. Explain to the students prior to beginning the test the nature and purpose of the object test.
2. Establish ground rules to be applied during the testing period, i.e., students will not confer with each other or share experiences.
3. Divide the class into groups having only as many taking the object test at one time as there are test stations. Assign other students class work, or possibly a paper-and-pencil test at this time.
4. Supervise the students taking the test and help students not able to follow instructions.
5. Inform the students of the time to change test stations following the predetermined plan of student rotation.
6. Do not permit a student to use more time than previously established for each station. Students should not be permitted to return to a test station after having left it.

ADVANTAGES OF THE OBJECT TEST

The object test provides a valid measure of the performance goals identified. Students recognize the fact that object tests are similar to actual performance activities required on the job; therefore, students usually accept these tests as being superior to paper-and-pencil tests. Object tests provide a realistic experience that in itself provides a valuable learning activity for students. Object tests are similar in nature to manipulative-performance tests; however, the main emphasis of object tests is identification. The identification is a preliminary step to application. Often application may be combined with identification in a limited manner in the object test.

Use of Manipulative-Performance Tests

If the student in courses of vocational and technical education is to be ready for employment upon completion of the educational program, it is desirable both for the student and the instructor that liberal sampling of his actual ability to perform job functions has been determined.

The best way to make such a determination is by giving effective manipulative-performance tests. The term "manipulative performance" is used to distinguish this type of test from paper-and-pencil tests. However, it is important to realize that mental processes, and often mathematical calculations, are significant elements of the manipulative-performance test. This is *not* just a test of the student's ability to push buttons or pull levers. In most instances the jobs of today in technical and trade and industrial fields require considerable depth in technology, applied mathematics, and sciences as well as the ability to think through a problem, organize the attack on the problem, and produce an effective solution.

NATURE OF THE MANIPULATIVE-PERFORMANCE TEST

A manipulative-performance test is a test designed to measure the student's ability to perform the selected operations of a specific job under rigidly controlled conditions. Implicit in this test are the possession of the ability to apply essential knowledges in addition to required manipulative skills. Even more significant is the ability of the student to meet established standards of quality and quantity while applying correctly principles and techniques of production or problem solving.

The instructor observes the student's performance of each step in the process and records on a prepared checklist the extent to which the student meets or fails to meet correct procedures and methods of performance. The time required by the student to perform each of the steps is indicated on the check sheet together with unique approaches or procedures which later may be discussed with the student. As a result of the manipulative-performance test a thorough review of the student's performance results. Further analysis of problems and difficulties becomes the basis for reteaching and other renewed learning experiences.

Manipulative-performance tests vary from very simple to extremely complex. Some tests require only a few minutes; others require several hours to complete. The instructor can build a manipulative-performance test of suitable difficulty and length by carefully developing the performance goals and selecting certain critical aspects of the job for emphasis.

BUILDING A MANIPULATIVE-PERFORMANCE TEST

To build an effective manipulative-performance test the following steps should be followed:

1. Determine the performance goal(s) which is to be evaluated.
2. Develop the specifications for each item of the manipulative-performance test.

3. Formulate the checklist and directions to be used by the person administering the test.
4. Establish the ground rules for the test. Clarify these for the participants. This includes such factors as time, equipment, materials, etc.

ELEMENTS TO BE RATED IN THE PERFORMANCE TEST

The selecting and weighing of the elements to be considered in the determination of performance are important. Usually these elements consist of:

1. Quality—production or service in conformance with a predetermined standard.
2. Quantity—production of a number of units within a specified time or the speed required to complete the job if only one unit is being produced.
3. Procedure—conformance with acceptable practices or given rules or procedures that are considered in the evaluation of the student's success. Observance of safety, care of tools and equipment, and similar factors are weighed in the final score assigned to the student for the completed job.

ADVANTAGES OF MANIPULATIVE-PERFORMANCE TESTS

Manipulative-performance tests offer several advantages. Some of the more important advantages are:

1. More valid measures of ability to do the job under actual job conditions are possible.
2. They are more objective than paper-and-pencil tests. Sometimes, individuals can verbalize well but cannot perform on the job; others perform effectively on the job but do not verbalize well.
3. Results of instruction are measured directly by the student's ability to perform. The instructor's accountability is more directly assessable. There is much truth in the saying "If the student hasn't learned the instructor hasn't taught."
4. Analysis can be made of the need for reteaching and relearning.
5. Morale of the student is improved. The relevancy of instruction is perceived by the student and translated into motivation and effort.

FACTORS TO CONTEND WITH

Instructors must realize the nature of manipulative-performance tests and recognize that:

1. More time is usually required to administer a manipulative-performance test than a paper-and-pencil test.

2. Consideration must be given to the size of the group which can participate in such a test at one time. Plans must be made for the students who are not participating.
3. Adequate planning must take place to provide effective control of the student's participatory efforts and the instructor's objective evaluation of the student's efforts.

ILLUSTRATIONS OF MANIPULATIVE-PERFORMANCE TESTS

The manipulative-performance tests shown in Exhibits 19–1 through 19–5 were developed by students in classes in performance evaluation.

(Text continued on page 303.)

EXHIBIT 19–1

PERFORMANCE TEST

NAME______________________________

DRAFTING—6 WEEKS TEST (75 points)
time: 50 minutes

Draw the front, top, and right side of the angle plate below.
Draw the views in the space below. Drawings should be *actual size.*
(Scale the drawing.)
(Do not dimension.)

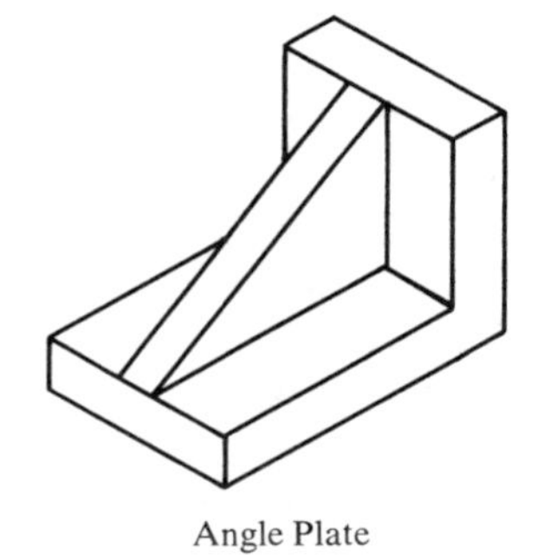

Angle Plate

EXHIBIT 19–2

MANIPULATIVE-PERFORMANCE TEST

COURSE AC LAB. ACTIVITY LAB. TEST SERIES CIRCUIT

DATE ______ TITLE PHASE ANGLE MEASUREMENT

NAME ______ GRADE ______

Materials needed:
1. 2-K resistor
2. 8-H inductor
3. 1-μF capacitor

Equipment needed:
1. Signal Generator
2. Circuit Patchboard
3. Clip Leads
4. Oscilloscope

Group 1—Set generator at 100 hertz and measure angle theta. Draw a circuit schematic and a pictorial representation of equipment connection.

ANGLE THETA = ______ degrees

Group 2—Set generator at 100 hertz and measure angle phi. Draw a circuit schematic and a pictorial representation of equipment connection.

ANGLE PHI = ______ degrees

When finished leave equipment connected and call for an instructor to grade the test.
Grading procedure: 50% angle measurement; 25% drawing; 25% bench neatness

EXHIBIT 19–3

Test of Manipulative Performance—Analog Computer

(Student directions)

Solve this set of simultaneous equations on the EC-1 computer.

Record the solution in standard lab note form.

$6X + 6Y = 6$ $5X - 3Y = 1$

(Points to check for evaluation)

Does student:

1. calculate proper feedback components. (4 points)
2. lay out proper schematic.
3. determine proper initial conditions. (2 points)
4. check on need for scaling.
5. hook up computer correctly. (5 points)
6. insert component carriers carefully.
7. allow for correct warm-up time.
8. use metering properly.
9. check amplifier balance just prior to running problem.
10. operate controls smoothly.
11. record results and procedure correctly.
    (A) define problem
    (B) show calculations
    (C) sketch hookup
    (D) identify component values
    (E) list sources of possible error
    (F) calculate probable accuracy
12. store equipment and components correctly.

EXHIBIT 19–4

Manipulative-Performance Test on Roof Framing Unit

Cut a common rafter out of 2 x 4 stock for a building with a span of 11 feet. The unit run is 12 inches and the unit rise is 4 inches. The rafter overhangs the building one foot and the ridge board is a 2 x 4.

Use the step method with the framing square to lay out all cuts. Make cuts with hand saw.

EXHIBIT 19–4 continued

PROCEDURE CHECKLIST
(Checked while student performs)

Item of Work	Maximum Credit	Credit Earned
1. Selects stock of appropriate length	2	
2. Checks that crown is up on stock	2	
3. Places square on stock using *4 on tongue* and *12 on body*	2	
4. Marks tail plumb cut, one step up, and marks building line plumb cut	2	
5. Plans five 12″ steps and one 6″ step to ridge plumb mark	2	
6. Shortens ¾″ for ridge	2	
7. Cuts birdsmouth 3½″ long	2	
8. Proceeds in a manner displaying knowledge and confidence	2	
9. Selects *crosscut saw* and makes plumb cuts	2	
10. Uses *rip saw* to cut seat cut	2	
11. Tests with square to see if saw cuts are square	2	
12. Exercises safety precautions in use of saw	2	
13. Finished product has total length of 81½ inches from ridge cut to end of tail	10	
14. Work performed in 20 minutes or less	6	
Total points	40	

EXHIBIT 19–5

OXYACETYLENE I
PUDDLING BEADS PERFORMANCE TEST

INSTRUCTOR'S DIRECTIONS

The instructor administering the test is to check in the column under yes or no if the student performed each step correctly. A place under each step is provided for the instructor's remarks to help in grading. The point value is indicated after each step. If the student correctly performs each step, he receives the point or points' value for that step.

EXHIBIT 19–5 continued

PUDDLING BEADS PERFORMANCE TEST

INSTRUCTOR'S CHECK SHEET

NAME________________________ APP. BY________________

SPECIMEN #__________________ SCORE_________________

DATE________________________ REMARKS_______________

____

STUDENT EVALUATION

			Yes	No
1. Student selects correct thickness of materials	(5 pt.)	Remarks	____	____
2. Makes layout lines for weld.	(5 pt.)	Remarks	____	____
3. Selects correct welding tip.	(5 pt.)	Remarks	____	____
4. Cleans tip correctly.	(5 pt.)	Remarks	____	____
5. Turns on oxyacetylene properly.	(5 pt.)	Remarks	____	____
6. Sets correct oxygen pressure.	(5 pt.)	Remarks	____	____
7. Sets correct acetylene pressure.	(5 pt.)	Remarks	____	____
8. Lights and adjusts torch properly.	(5 pt.)	Remarks	____	____
9. Holds the correct torch angle (45 degrees to 60 degrees).	(5 pt.)	Remarks	____	____
10. Holds proper distance from tip to base metal.	(5 pt.)	Remarks	____	____

EXHIBIT 19–5 continued

SPECIMEN EVALUATION OF BEAD

11. Width specified. (10 pt.) Remarks ______

12. Straightness specified. (10 pt.) Remarks ______

13. Penetration specified. (10 pt.) Remarks ______

14. Appearance specified. (10 pt.) Remarks ______

15. Length specified. (10 pt.) Remarks ______

PUDDLING BEADS PERFORMANCE TEST

STUDENT'S INSTRUCTION SHEET

You are to use the oxyacetylene welding outfit and run three beads as indicated on next page. The tolerance of specifications is described under the Standard of Acceptability.

GENERAL INSTRUCTIONS

A. Select correct material as indicated.
B. Lay out lines for beads.
C. Select welding tip.
D. Turn on oxyacetylene outfit.
E. Proceed with work.

TEST SPECIMEN DETAILS: MILD STEEL (ASTM A-7)

DIRECTIONS:

Use chalk to lay out a line for weld.

Bead #1 is 4 inches long
Bead #2 is 5 inches long
Bead #3 is 2 inches, 1 inch, 3 inches

EXHIBIT 19–5 continued

6″

Bead #1

Bead #2

Bead #3

90°

90°

4″

STANDARD OF ACCEPTABILITY: Show to instructor for grade.

(a) Bead width—Must be $\frac{5}{16}$ in. wide within $\pm\frac{1}{32}$ in.
(b) Bead straightness—Must be straight within $\frac{1}{16}$ in. of center axis from start to finish.
(c) Bead penetration—Bead must *not* penetrate through on bottom side.
(d) Bead appearance—Must have even, finely spaced ripples.

W.N.V.T.S. OXYACETYLENE I CHALK INITIALS ON EACH SPECIMEN AT XXX

Subject: PUDDLING BEADS PERFORMANCE TEST

NAME ________ APP. BY ________

DATE ________

Bead #1 ________

SCORE: ________ Bead #2 ________

REMARKS: Bead #3 ________

Summary

Object tests provide a valid measure of performance goals based on identification and selection of materials, tools, equipment, etc. Careful planning and administering help to assure effective evaluation.

Manipulative-performance tests are excellent for evaluation of the student's ability to perform on-the-job functions under actual job conditions. This test may indicate the need for reteaching of concepts and skills previously covered.

In building a manipulative-performance test:

1. Determine the performance goal(s).
2. Develop the specifications.
3. Establish the checklist and directions.
4. Make the rules for the test.

Usually, in the manipulative-performance test consideration is given to production quality, quantity, and procedure.

Manipulative-performance tests for shop or laboratory work are likely to be valid and objective measures of competency on the job. The relevancy is so apparent to most students that students' morale and motivation are elevated.

QUESTIONS AND ACTIVITIES

1. Explain the differences between a paper-and-pencil test and an object test, and a manipulative-performance test.
2. When should the object test be used? Why?
3. When should the manipulative-performance test be used? Why?
4. List the steps in the construction of a manipulative-performance test. Of an object test.
5. Outline the procedure to be followed in administering each of these types of tests.
6. What factors are considered in rating the manipulative-performance test?
7. Discuss the disadvantages of the manipulative-performance test.
8. Plan and develop an object test for a course which you teach now or may teach in the future.

9. Plan and construct a manipulative-performance test for a subject you now teach or may teach in the future. Give:
   A. Performance goals
   B. Test items
   C. Procedure for administration
   D. Procedure for evaluation

Chapter 20 OBSERVING STUDENTS AT WORK AS A BASIS FOR EVALUATION

Observation is a useful tool in evaluation. Through observation impressions are developed which often constitute the basis for an informal assessment. Frequently, this informal assessment is one of the most significant resources for appraisal used in the completion of confidential papers relative to the student's competency, oral recommendations, and rating forms (i.e., rating scales). It has often been said that the prospective employer places more confidence in such recommendations than in the interpretations of formal grades. Observation may be a substantive aid in arriving at an equitable formal grade also.

Some teachers have said that after working with and observing students at work an equitable grade can be determined without the use of formal tests of any kind. While this may be carrying the powers of observation further than most educational systems are willing to permit, it certainly does indicate the confidence some people place in powers of observation. This reflects to some extent the attitude of administrators in business and industry.

Observation and the impressions created by the individual together with his productive accomplishments are major factors in rating in most industrial firms. Success in industry for the individual requires a satisfactory quantity of acceptable units of production (or service). But, the supervisor's observation of the employee's attitude, appearance, use of tools, equipment, and materials, manner of working, associations with other individuals, and conscious efforts to perform effectively are important weighing factors in the final assessment given most employees by many employers.

Observation, then, has a legitimate role in the assessment of students in school, since experiences in school should mirror, to some extent at least, the conditions that the student will find later in the world of work.

Weaknesses of Observation

If observation is to serve as a means of arriving at a fair evaluation of the performance and the performance potential of the student, the same basic factors must be considered as were identified as the criteria for building any other type of test.

Observation may not be as effective and as equitable as it possibly could be because the observer does not apply the same conscious effort as is used in building a paper-and-pencil test or a manipulative-performance test.

Bias may be introduced into the evaluative decision due to personal elements present in the observer, such as how the observer feels, his involvements with people (i.e., an argument with another employee, a bad conference with the boss, a problem at home), his reaction to prejudicial elements of the environment (i.e., the worker wears long hair and the observer detests long hair), or previous contacts with the individual being observed (he has had words with the employee before).

Bias may also be introduced into the evaluative decision because of failure to use adequate time to see performance over a sufficient length of time to correctly weigh the performance in perspective to the objectives to be accomplished. While a keen observer may through a "spot checking" or sampling approach get some reaction relative to the broad approach of the individual to his job, many deductions may be "read in" or "read out" which should be considered for a fair evaluation of performance.

Bias may be introduced into the evaluation due to failure to record observations. Memory may be relied upon without recording the observations. Time erases many experiences from memory. A written report in the form of an anecdotal record, observation check sheet, or rating scale is desirable to make observations more objective.

Observation, then, is subject to many possible errors and biases. However, since it is one of the most frequently used methods of rating employees, it does have a place among the evaluative techniques used in school. But, the raters need to be aware of the weakness of observation as an evaluative tool.

Criteria for Effective Observation

Effective observation as a good test must conform to the following elements of criteria:

1. Validity	observers must see those elements that influence the success of the worker on the job and eliminate those elements that do not.
2. Reliability	observers must consistently observe the important elements of the criteria without being influenced by personal experiences or environmental elements which tend to distort judgment.
3. Objectivity	observers must be fair to students.
4. Comprehensiveness	observations must be long enough to be effective.
5. Ease of recording	a check sheet or other record form should be used to provide a permanent record of sampled observations. A fair plan of sampling observations needs to be used. Reference to the written record needs to be made prior to formulating the final evaluation, rating score, or grade.

In some cases failure to apply valid criteria for observation has reduced observation to identification of the most flagrant violations of good practice; while in other situations every detail of the student's errors was noted. This may in reality be a reflection of the observer's likes and dislikes for the individuals observed. Many people see only that which they want to see.

Observation is a good tool for evaluation if used properly. Application of the five criteria just listed will help to ensure proper use.

RECORDING OBSERVATIONS

Observations should be recorded as soon as possible after having been made. Individuals making observations frequently prefer to record these observations after rather than while visiting the student.

Anecdotal records, consisting of written paragraphs relating the elements observed, may be used for this purpose. These records may contain written accounts of pertinent actions or conversations of the individual. Rather than lengthy written discourses these notations may be reduced to meaningful, brief sentences of the most pertinent observations. Anecdotal records have been used to build case histories in situations where difficult problems require validating information to assist in arriving at solutions. Anecdotal records are helpful, but do require considerable time if several students are involved.

Check Sheets and Rating Scales

Check sheets prepared with spaces for checking off the effectiveness of the student in the performance of the assigned tasks are much faster and more likely to be reliable. The pertinent items are identified in advance with space for adding additional details. The check sheet may be constructed to provide a series of degrees of effectiveness. A three-point scale or a five-point scale may be used for this purpose as well as a simple yes–no system.

A rating scale may be developed which will permit the observer to indicate values at any interval on the scale for each item from the highest possible value to the lowest available.

Discretion needs to be used in selecting the items to be included in the rating form. A simple form, covering the main concerns, is usually adequate and much easier to use. A complicated, time-consuming form may defeat the purpose and not be used at all.

An illustration of a check sheet used in observing student teachers is shown in Exhibit 20–1.

EXHIBIT 20–1

EVALUATION (OR SELF-EVALUATION)

DIRECTIONS: Determine and circle the most accurate score of the scale. The highest value "5" corresponds to an "A" with the lowest value "1" corresponding to "F." Then list significant comments concerning the strengths and weaknesses of the individual's performance as a teacher.

STUDENT TEACHER____________________DATE________

SUPERVISING TEACHER________________SCHOOL________

RATING FACTORS	DESCRIPTION OF RATING FACTORS	SCALE 1 2 3 4 5
1. Work habits	Promptness, efficiency, cooperation, safety, etc.	. . . . .
2. Working with people	Ability to get along with students, teachers, and others	. . . . .
3. Preparation	Effective planning and organization of work	. . . . .
4. Control of student interests	Ability to motivate and stimulate students to constructive activity and progress	. . . . .
5. Teaching procedures	Application of the teaching processes: demonstration, discussion, problem solving, etc.	. . . . .
6. Teaching activities	Involvement of students in doing activities meaningful and useful to goals established	. . . . .
7. Student evaluation	Follow-up activities, reteaching if necessary, testing, and evaluation of student learning	. . . . .
8. Community and/or school participation	Mixing with associates; participation in school–community activities	. . . . .
9. Personality for teaching	Effective, understanding, and friendly but not familiar with students	. . . . .
10. Subject matter competencies	Adequate for units taught with ability to provide useful illustrations and experiences	. . . . .
11. Assignments	Clear, definite, and within ability of students	. . . . .
12. Willingness to learn	Willing to assume responsibility and accept suggestions; effective use of time and energy	. . . . .

EXHIBIT 20–1 continued

13. Personal characteristics	Attitude, appearance, initiative, honesty, leadership, and other qualities desirable for teachers	. . . . .
14. Use of teaching aids	Uses effectively text and reference books, study guides, blackboard, charts, motion pictures, etc.	. . . . .
15. Plans and reports	Prepares requested plans, reports, and other similar materials promptly and accurately	. . . . .
16. Care of facilities	Applies good housekeeping and care of room	. . . . .

COMPOSITE RATING—evaluation of student teaching reduced to a single value . . . . .

COMMENTS ON STUDENT TEACHING

STRONG POINTS	POINTS THAT NEED STRENGTHENING

Reviewing Observations

While observations and other evaluative devices may be used for multiple purposes, their two main justifications in education are the improvement of instruction and the grading of the student.

If the results of observation are to produce improved performance by the student, it is essential that the instructor review with the student the strengths and weaknesses of his (or her) present level of performance.

Accentuating the positive is the application of good learning psychology. After pointing out a few of the strong points in the student's work or behavior, it is desirable to identify the elements that need to be improved. Suggest in a friendly way not only what the student is doing wrong, but how he should correct errors or strengthen and improve present methods of performance. Properly done this will result in correction of habits and improvements in attitude of the student. The teacher needs to demonstrate understanding and awareness of the student's problems. As the teacher builds a positive attitude and demonstrates empathy the climate changes, permitting the teacher to reach the students more effectively. Reinforcement of learning coupled with practice under effective supervision aids in the development of confidence, the desire to achieve more effectively, and the improved quality and quantity of production.

If the student has many problems, select a few of the most essential for immediate emphasis and attention. Help to create an environment which results in success rather than failure. "Nothing succeeds like success." Build the reputation of the student as needed, little by little, and he will grow and improve.

Record information as you observe the growth and improvement of the student. The next step is to translate the impact of student learning as well as the earlier observations into units which can be combined with other units to produce a score reflecting the total evaluation in the form of a final grade.

CONVERTING OBSERVATIONS INTO MEASURABLE UNITS

If the results of observation are to be fairly combined with other data compiled relative to the performance of the student, it is helpful to devise a plan built upon numerical values.

The teacher may decide that the components of the final grade shall consist of items such as the following:

Performance on written tests.
Performance on manipulative-performance tests.
Completed projects—number and rated quality (in points).

Observed performance—shop and/or laboratory work.

It is then necessary to translate each of the above into the appropriate percentage of the total value. This, then, makes it possible to evolve a plan for weighing each of the sampled observations as recorded throughout the marking period.

PHILOSOPHY FOR OBSERVERS

Observers need to cultivate a critical attitude in harmony with the performance goals defined, if observation is to be a useful instructional and evaluative tool. A careful and methodical approach is desired. The observer must be aware of different levels of acceptable performance as well as the unacceptable level of performance.

The teacher must not be influenced by previous levels of accomplishment by the student. Each element of each observation needs to be rated independently of the previous achievements of the student.

Significant elements of the observation must be weighed in perspective to the total objectives of achievement sought. The instructor must know from personal experience what the essential behaviors and "earmarks" of successful performance are for that occupation and the level of proficiency required by the user of the product—the employer.

The emphasis of observation must be on the student's application of salable skills, knowledges, habits, and attitudes.

Summary

Observation is probably the most commonly used evaluative device in industry and business. Observation can be a helpful addition to evaluation of the student's performance in school.

If observation is to be used as a basis for evaluation every effort must be made to be objective. Bias must be minimized and errors of observation eliminated or reduced to a low level.

Every effort must be made to make observations:

1. Valid
2. Reliable
3. Objective
4. Comprehensive

Observations may be recorded using anecdotal records, check sheets, or rating scales.

Observations must be translated into the equitable values for combination with other elements of the total evaluation for the final grade. This can be achieved most easily through the use of numerical values. Proportionate weights must be assigned to each observation.

The focus of observation must be on the student's application of salable skills, knowledges, habits, and attitudes.

QUESTIONS AND ACTIVITIES

1. Is observation a realistic element of the total system of evaluation? Justify your position.
2. What must be done to make observations:
   A. Valid?
   B. Reliable?
   C. Objective?
   D. Comprehensive?
   E. Accurately recorded?
3. If observations are suitable for promotion and retention of employees in industry and business, why should their use in evaluating students be less desirable?
4. How can bias be reduced in observation for evaluation?
5. What is the role of interpretation in observation? Of records?
6. Explain the method of using anecdotal records.
7. What are the advantages of using check sheets? Disadvantages?
8. What are the strengths and weaknesses of rating scales for recording observations?
9. Why should the results of observations be reduced to numerical values?
10. What concepts are vital to a philosophy of observation?
11. Develop a plan for using observation as an element of student evaluation for purposes of grading.
12. Keep an anecdotal record of your students' critical incidents for a two-week period. Evaluate the results.
13. Develop a check sheet for recording elements of performance vital to evaluation.
14. Develop a rating scale for observation suitable for evaluation of observations.
15. How do you translate observations into final grades for the students observed?

Chapter 21 CONVERTING THE STUDENTS' ACHIEVEMENT TO GRADES

The American Society for Engineering Education in the *ERM Newsletter* suggested that the computer may assume a significant role in the future in evaluation. In this issue Les Harrisberger said in part:

> . . . It may be that one day it will be possible for each student to report to a console at a time of his own choosing and take an examination over the material he is currently studying. I can envision what I would call a dynamic response exam; that is, the computer would prepare the second question on the basis of the answer received for the first. No student would get the same examination since the examination would be constructed according to student response. The computer would evaluate each of the responses and issue a report that would be a relative measure of the ability and progress of that student compared to all the others who took the test. A summary evaluation of the student's performance would be issued to the student and also automatically logged in the student's master file and in the course instructor's grade file. (1, p. 2)

While most teachers would undoubtedly welcome such an innovative solution to the process of converting students' achievement into grades, this at present seems a long way away.

Sources of Information for Determining Grades

The foundation for determining the students' grades was built at the time that the initial course of study was constructed. At that time the criterion objectives were determined. The criterion objective is a terminal, end-of-course objective, which establishes the behavior that the student will be able to demonstrate at the end of the course.

One of the elements of a performance goal (behavioral objective) is behavior (visible activity of the learner); the other is criterion (standard for evaluating terminal behavior). The criterion is usually expressed either in terms of quality or quantity, and sometimes both.

The criterion objectives need to be coupled with evaluative items or devices to provide information for determination of performance. Evaluation of the performance of the students on the criterion objectives may be reflected in many ways and through a variety of instruments, including:

1. paper-and-pencil tests
2. object tests
3. manipulative-performance tests
4. finished product
5. observation

The results of all evaluative findings must be brought together and reflected equitably in the final grade for the course. This final grade needs to convey to prospective employers and students alike a realistic and fair assessment of the performance capability of the students at the time the course of instruction is completed.

The teacher should develop for each course a table of specifications which identifies the relative value of each evaluative method used for determining the final grade. Since numbers are much easier to work with than letters, it is suggested that the results of each evaluative method be reflected in numerical values during the evaluative process and translated into the kind of unit required for the final grade after all the raw data have been collected. This may be as follows:

Paper-and-pencil tests	200 points
Manipulative-performance tests	200 points
Finished product	200 points
Observation of students	100 points
Total	700 points

In this illustration the highest possible scholastic achievement would be reflected by the students with the largest number of points.

Use of Absolute or Relative Criterion

At this point, comparison of the achievement of each student is possible. If an absolute criterion is used the comparison is against a predetermined standard. Success on the job in vocational and technical education is reflected by quality and quantity of production. If the quality or quantity of production falls below a predetermined standard the position of the individual may be in jeopardy. If this concept is to be translated into the learning process the instructor must validate in terms of employment standards the standards to be used for evaluation of the students' performance in school. The result might be that for students to be employable in the occupation they must achieve a performance level of __ percent. Study of the individual occupation and the level of employment will help in realistically setting the standard of performance. It may be that the decision is reached that each individual needs to achieve at the level of 60 percent to qualify for employment in the job market. Some will achieve at levels much higher. It is therefore necessary to translate achievements of higher than 60 percent into grades that reflect a higher performance of acceptable achievement during the learning period.

The marking scale adopted may be as follows:

90–100% = A
80– 89% = B
70– 79% = C
60– 69% = D
Below 60% = F

Industry, business, or agriculture may in some instances require an extremely high standard of performance in relation to certain specific activities. Where safety is involved anything less than 100 percent may be unsatisfactory. In other cases, such as production for the space industry, extremely high standards and careful determination that assurance of quality has been achieved must be required. These types of employment conditions suggest the value of using a rigid absolute criterion in some situations in the educational process.

During recent years the academic community has placed considerable emphasis on relative criterion as the basis for determination of the evaluation of performance. The achievement of each student has been viewed in terms of the achievement of other students in the class. Statistical methods have been employed to determine the relative achievement of one student in terms of the achievement of the other students in that class or possibly in several classes conforming to the same evaluative instruments and evaluative criterion. Measures of central tendency (mean, median, or mode) have been used together with measures of dispersion (standard deviation or average deviation). Then, results have been translated into letter grades for the students.

Correct results using the relative criterion approach require exercise of precaution relative to randomization and the selection process. Beginning with the elementary grades some selection has taken place. The higher in school students progress the more selection has taken place. If this concept is applied at the college or university level a great deal of selection must have taken place.

It is conceivable that because of the selection process all individuals, even those who failed, may have achieved at a level adequate to meet the requirements of the employer. It is likewise possible, using a relative criterion, that individuals receiving the highest grade would still *not* have achieved at the level required by the employer for a particular position.

The standard, absolute or relative, must be determined by the instructor in conformance with the criterion objectives identified and with established school policy.

THE CASE FOR USING ABSOLUTE CRITERION

One of the major curriculum objectives of all vocational and technical education is employment in the occupation upon completion of the educational program. While industry, business, and agriculture use individuals of varying competency and this variation exists from firm to firm, determination of the minimum acceptable skills, knowledges, habits, and attitudes is an important aspect of occupational analysis. Once this has been determined it is then necessary to transfer this standard to the educational program. Taxpayers will not long support vocational and technical programs of education that do not meet this standard. Achievement equal to or in excess of this standard becomes a "must."

There are some who argue that pass–fail grading is the most realistic system. They say that in industry either the job is acceptable or it is not acceptable. Hence, the pass–fail system of grading is similar to the practice on the job. Others are quick to point out that in school grades provide a motivating effect which challenges students to achieve at a high rate or with superior quality. This seems to be one of the justifications for having two, three, or four degrees of acceptable work reflecting variations in achievement and corresponding letter grades.

The setting of the absolute criterion is a most important step in the total educational process under this approach. The judgment of the instructor needs to be supported by validation of both curriculum content and standards of performance. Use of the consulting curriculum committee, information questionnaire to industry, business, or agriculture, and on-site job analysis is highly desirable to provide the basis for the standard. Periodic adjustment must occur to reflect changes in technology and the emphasis of the user of the individuals who completed the educational program.

The realism and practicality of the absolute criterion, when correctly established, provide a strong case for this approach. They also present a personal challenge to each student to achieve or exceed the criterion required for employment. The competition is not with the peer group, it is against a standard of the industry essential for employment.

THE APPLICATION OF THE RELATIVE CRITERION

If a relative criterion is to be used the main source of information for the standard will be the performance of the class or whatever total group is used for comparison. If a large number of samples is taken from a specified population the means of the sample will tend to be normally distributed. Some other statistics also tend to be normally distributed. A normal distribution makes up a bell-shaped curve graphically illustrated in terms of the mean and standard deviation. By fixing the mean at the midpoint of the bell-shaped curve and then establishing the standard deviation, grades can be assigned to individual cases as they fall within the area of the curve.

The total area under the curve is equal to 100 percent. As shown in Figure 21–1, in a normal distribution the spread of students' grades is:

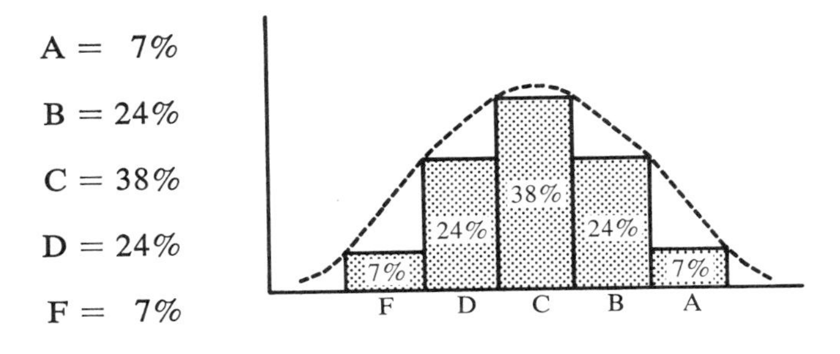

FIGURE 21–1

BELL-SHAPED CURVE

The comparison of one student's achievement with that of the others in the total group is achieved by the use of statistical measures such as the mean and the standard deviation.

The number of cases should be approximately 100 or more if the relative criterion is to be a valid basis for regular statistical treatment.

In some cases other statistical treatment may be employed with small numbers of cases that will be adequate.

GATHERING THE RAW DATA FOR USE WITH RELATIVE CRITERION

The process of gathering the required raw data is basically the same with the relative criterion as with the absolute criterion. The decision must be made as to what will be considered in the evaluation of performance of the individual. Rarely in vocational and technical education is a final grade based upon a single test. It is therefore necessary to establish the basis for evaluation of performance as suggested earlier in this chapter. Again, it is highly desirable to retain the values in numerical form until the final determination of grade is to be made.

It is now assumed that the total points earned by each of the students on the various elements of the criteria (i.e. paper-and-pencil tests, manipulative-performance tests, finished product, and observation) have been compiled.

Translation of the raw data into meaningful terms can be achieved by:

1. Arranging the scores in descending order to provide the range of scores.
2. Determining the central tendency (mean, median, or mode).
3. Determining the deviation (average deviation, standard deviation).
4. Assigning marks on the basis of a normal distribution using the means and standard deviation.

While various methods may be used to complete the statistical process the classroom teacher usually needs to employ only the basic manual mathematical calculations. Different formulas are available for grouped data and applications to calculators are provided in most textbooks on statistics. In this text the fundamental processes only will be presented.

Frequency Distribution

A frequency distribution shows how the scores are distributed. A frequency distribution not only indicates the range of scores but it also gives the number of times the various scores occurred. Placing the scores in a frequency distribution helps to provide additional information for interpreting the scores.

When the scores are close together they may be arranged with intervals of 1; whereas, when the scores are more widely separated the interval may be 2, 3, 5, or even 10.

The process of building a frequency distribution consists of listing scores from highest to lowest and tallying the number of times that any of the scores are repeated. This can be built into a table (see Exhibit 21–1).

EXHIBIT 21–1

FREQUENCY DISTRIBUTION

Scores	*Tally Marks*	*Frequencies*
95	/	1
94	/	1
93	//	2
92	/	1
91	/	1
90	///	3
89	////	4
88	~~////~~	5
87	~~////~~///	8
86	////	4
85	~~////~~////	9
84	~~////~~	5
83	///	3
82	//	2
81	/	1
	Total Scores	50

RANGE

The range of scores is simply the difference between the highest and the lowest scores. In the illustration of frequency distribution (Exhibit 21–1) the range is 14 since the highest score was 95 and the lowest score was 81.

MEASURE OF CENTRAL TENDENCY

A measure of central tendency is an average. The variation of raw scores is reduced to a single value. There are three kinds of averages. These are the mean, the median, and the mode.

Mean. The mean is the same as the common average. If the number of scores is small or if an adding machine is available, the mean can be found by simply adding all the scores and dividing by the number of scores. If four students earned scores of 90, 80, 70, and 60 the total of the four scores, 300, divided by 4 gives the mean of 75.

When the number of scores is larger a method of finding the mean by the use of a frequency distribution may be preferred. To establish the size of interval to use, first determine the range. Divide the range by a number that will result in about ten intervals or somewhere between ten and twenty intervals. Then determine the size of the intervals and set the limits of the intervals. Arrange the intervals in order from highest to lowest in a vertical column. Tally the scores and determine the frequencies for each of the intervals. It is then necessary to place the values in a frequency table and apply the following formula:

$$\text{Mean} = \text{Assumed mean} \times \frac{\text{Sum of frequency} \times \text{deviations}}{\text{number}} \times \text{Interval}$$

The process is shown in Exhibit 21–2.

EXHIBIT 21–2

Calculation of Mean

STEP I	*STEP II*	*STEP III*	*STEP IV*	*STEP V*	*STEP VI*
Establish the intervals	Tally scores	List frequencies	Determine the AM	Record deviations	Multiply frequency x deviation
30–34	/	1	(assumed	3	3
25–29	//	2	mean). In this case	2	4
20–24	//	2	it is in	1	2 (+9)
15–19	𝍸	5	the 15–19 interval.	0	0
10–14	//	2	In this	−1	−2
5– 9	/	1	case	−2	−2
0– 4	/	1	AM = 17.5.	−3	−3 (−7)
(Interval of 5)		N = 14			(Efd = 9 − 7 or 2)

Application of the formula gives the following:

1. FORMULA $M = AM + \frac{Efd}{N} \times i$
2. $17.5 + \frac{2}{14} \times 5$
3. $\frac{2}{14} \times 5 = \frac{10}{14}$ or .71
4. $.71 + 17.5 = 18.21$
5. The mean is 18.21

Median. The median is the 50th percentile. It is the value above which and below which 50 percent of the scores lie. A simple way to find the approximate median is to arrange the scores in order from highest to lowest. The middle score represents the median.

The median can be calculated from a frequency distribution. The procedure consists of the following steps:

1. Place the scores in a frequency distribution.
2. Locate the interval containing the middle score.
3. Add the scores cumulatively starting from the bottom recording the sum at each interval.
4. Add to the lower limit of the interval in which the median falls that proportion of the amount of this interval needed to reach the exact midpoint.

The process is shown in Exhibit 21–3.

EXHIBIT 21–3

CALCULATION OF MEDIAN

STEP I	*STEP II*	*STEP III*	*STEP IV*	*STEP V*
Place in frequency distribution	Tally the scores	Total the scores	Add the scores cumulatively	Locate the median interval
90–99	/	1	20	N/2 or
80–89	//	2	19	20/2 = 10 10th score
70–79	/	1	17	is in the
60–69	////	4	16	interval
50–59	~~////~~	5	12	50–59
40–49	///	3	7	
30–39	/	1	4	
20–29	//	2	3	Apply for-
10–19	0	0	1	mula
0– 9	/	1	1	
(In this case use interval of 10)		(Total = 20)		

STEP VI

Apply formula

Median is equal to lower limit (L) of interval containing desired point plus ½ cases (N) − cumulative frequency below interval (fb) divided by frequency (f) in interval times size of interval (i) used.

$$Md = L + \frac{\frac{1}{2}N - fb}{f} \times i$$

$$Md = 50 + \frac{10 - 7}{12} \times 10$$

$$Md = 50 + \frac{3}{12} \times 10$$

$$Md = 50 + 2\frac{1}{2} \text{ or } 52\frac{1}{2}$$

Mode. The mode is the score in the distribution which occurs most frequently. The mode is the least commonly used measure of central tendency. Some groups or distributions have two modes in which case it is described as bimodal. The mode can be easily determined by inspection. It is of little value except to suggest the approximate region of the mean.

Given the scores 90, 82, 80, 74, 72, 72, 72, 69, 68, 61, 58, the mode is 72.

The above measures of central tendency describe the location of the averages in a distribution of scores.

MEASURES OF VARIABILITY

If statistical treatment is to be applied to determine students' grades it is not enough to calculate the measure of central tendency. The spread of the scores must also be calculated.

To determine the scatter of the scores either the average deviation or the standard deviation is used.

Average deviation. The average deviation is the mean of the deviations of the scores from the mean. The process is that of listing the scores, finding the mean, determining the deviation of each score from the mean, and, then, the total of the deviations. Divide this value by the number of numbers.

Standard deviation. The standard deviation is the most accurate measure of variability and is most frequently used. It is defined as the square root of the average of the squares of the individual deviations from the mean. Sigma (σ) is the Greek letter used to indicate standard deviation.

The first two steps in finding the standard deviation are the same as for finding the average deviation. That is: find the mean and then find the deviation from the mean of each score. The next step is to square the deviation from the mean. Then find the sum of the squared deviations. Finally, substitute the values in the formula:

$$\sigma = \sqrt{\frac{Ex^2}{N}}$$

A simple method of calculating the standard deviation is to find the sum of the high sixth of the scores minus the sum of the low sixth of the scores divided by half the scores. The process is:

1. Add the number of scores.
2. Divide the number of scores by six.
3. Add the highest one-sixth of the scores.
4. Add the lowest one-sixth of the scores.
5. Subtract the low sum from the high sum.
6. Divide the difference by one-half the number of cases. (2, p. 19)

The calculation of standard deviation by the simple method is illustrated in Exhibit 21–4.

EXHIBIT 21-4

CALCULATION OF STANDARD DEVIATION

Suppose 30 students completed a 20-item test with the following scores:

Item		*Cases*	*Item*		*Cases*	*Item*		*Cases*
20	—	1	12	—	3	6	—	0
19	—	2	11	—	2	5	—	2
18	—	3	10	—	1	4	—	1
17	—	0	9	—	1	2	—	0
16	—	5	8	—	1	1	—	3
14	—	2	7	—	2	0	—	0
13	—	1						

Steps to follow:

1. 1/6 of 30 students is 5 students.
2. Sum of the 5 highest scores: 20, 19, 19, 18, and 18 = 96.
3. Sum of the 5 lowest scores: 1, 1, 1, 4, 5 = 12.
4. 96 − 12 = 84
5. 84 ÷ 15 = 5 3/5—the standard deviation.

NOTE: 15 is one-half the number of cases.

Assignment of Grades—Absolute Criterion

A class in manufacturing processes consisted of 20 students. The highest possible score on all criteria for performance evaluation was 500 points. Total scores earned by the students were as follows:

Accumulated Score	*Number of cases*
495	1
485	1
480	1
460	1
450	1
440	1
425	2
415	2
405	1
390	2
385	3
375	2
360	2

If a 70 percent absolute criterion were used the following grade distribution would be realistic (if the standard provided for three levels of passing work and one level of failing work).

Grade	*Score*	*Cases*
A	450–500	5
B	400–449	6
C	350–399	9
F	Below 349	0

If the nature of the course content were such that to properly prepare individuals for entry level in industry individuals must achieve 80 percent or better (absolute criterion of 80 percent), the results might be as follows if a pass—no pass grading system were being used:

Grade	*Score*	*Cases*
Pass	400–500	11
No pass	Below 399	9

The most frequently used absolute criterion system has four degrees of successful accomplishment and one indicating failure. In such a situation the allocation of grades would be as follows, using a 60 percent criterion standard:

Grade	*Score*	*Cases*
A	450–500	5
B	400–449	6
C	350–399	9
D	300–349	0
F	Below 299	0

Using an absolute criterion, careful consideration must be given as to where the standard of acceptable work is to be set and how many degrees of acceptable work will be employed. Inherent in such a system is the problem of determining the elements which shall constitute the components for evaluation of performance (paper-and-pencil tests, manipulative-performance tests, observation, finished project, etc.) and what relative emphasis is to be given to each. This same problem exists, however, if a relative criterion is used. The most realistic references are the requirements of the prospective employers on entry-level jobs. Knowledges and skills for meeting this standard may be grouped as follows:

Knowledge and skills	*Essential for employment*
Must know and be able to do	Cannot be employed without competency
Should know and be able to do	Desirable for employment
Nice-to-know and be able to do	Adds to value of employee for employment

Assignment of Grades—Relative Criterion

Considering the same class in manufacturing process of 20 students the steps in arriving at grades using a relative criterion might be:

1. Determine the range of scores
   495 − 360 = 135, the range
2. Find the mean (sum of all scores divided by the number of scores).
   In this case the simplest procedure is to add the scores on an adding machine and divide by 20 (8300 ÷ 20 = 415). Mean is 415.
   Note: If a frequency distribution were used with a range of 135 it would be convenient to use an interval of 10.
3. Find the standard deviation (square root of the average of the squares of the individual deviations from the mean). Either the formula or the abbreviated method would be satisfactory. If the abbreviated method is used, the process consists of:
   a. Find $\frac{1}{6}$ of the number of scores ($\frac{1}{6} \times 20 = 3\frac{2}{3}$)
   b. Add the highest $\frac{1}{6}$ of scores (495 plus 485 plus 480 plus 306 = 1766). Sum of $3\frac{2}{3}$ highest scores
   c. Add the lowest $\frac{1}{6}$ of scores (360 plus 360 plus 375 plus 250 = 1345). Sum of $3\frac{2}{3}$ lowest scores
   d. Subtract the sum of the low from (1766 − 1345 = 421) the sum of the high
   e. Divide the difference by one-half 421 ÷ 10 = 42.1
   f. The standard deviation is 42.1
4. Using the mean (415) and the standard deviation (42.1) apply the distribution of the normal curve as shown in Figure 21–2 to the data of the problem.
5. Translate the scores earned by the students of the class into grades using the mean (415), standard deviation (42.1), and the normal distribution. The logical assignment of grades in this illustration is given in Exhibit 21–5.

Summary

Grades should be established in conformance with the criterion objectives established. Criterion objectives are the terminal or end-of-course objectives. The criterion objectives should be performance oriented.

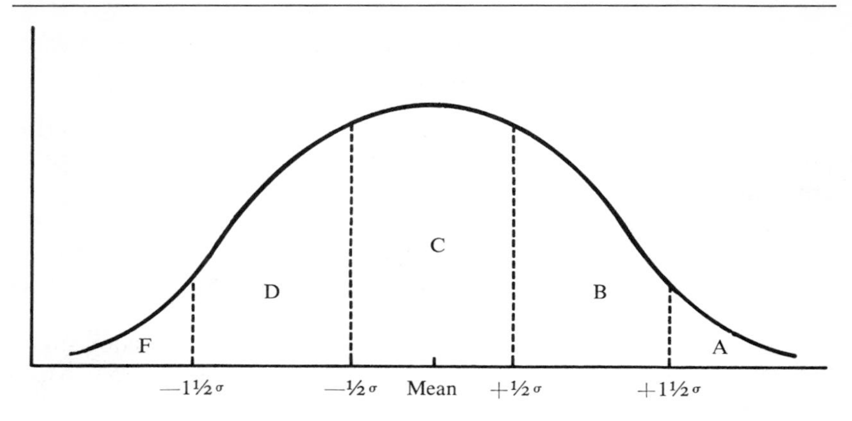

DISTRIBUTION OF NORMAL CURVE

FIGURE 21–2

EXHIBIT 21–5

ASSIGNMENT OF GRADES

Mean of group			415		
Standard deviation		$-1\frac{1}{2}\sigma$	$-\frac{1}{2}\sigma$ and $+\frac{1}{2}\sigma$	$+1\frac{1}{2}\sigma$	
Standard deviation applied values	000–351	352–393	394–436	437–478	479–500
Grades of group	F	D	C	B	A
Number of cases (Students)	0	9	5	3	3

Note: The highest and lowest possible values of the A's and F's are the maximum and minimum possible scores.

The results of all evaluative findings should be brought together in the final grade in proportion to a predetermined plan.

An absolute criterion uses a predetermined standard of performance. In vocational education this should be geared to the achievement level essential for employment on entry-level jobs of the specialty.

When a relative criterion is used the achievement of each student is viewed in terms of the achievements of the other students in the group. Relative criterion *does not* ensure adequate preparation for entry-level jobs.

The relative criterion is used with a normal distribution. The means and standard deviation are calculated and used in determining grades in proportion to the normal distribution as shown by the bell-shaped curve. The spread of students' grades is

A = 7% D = 24%
B = 24% F = 7%
C = 38%

When the absolute criterion is used the standard may be achieved by all, part, or none of the students. Those who achieve or exceed the standard should be employable.

Measures of central tendency include the mean, median, and mode. Measures of variability are the average and standard deviation.

QUESTIONS AND ACTIVITIES

1. What is the difference between relative and absolute criterion?
2. When should the absolute criterion be used? The relative criterion?
3. What is a terminal objective?
4. What are the two basic elements of a performance goal?
5. Explain the advantages of retaining and accumulating numerical data for each evaluative instrument and translating this into a grade after all the data have been compiled.
6. How should the performance level for satisfactory performance be determined?
7. What is a normal distribution?
8. How can the spread of grades for students be justified using a normal distribution?
9. Explain the use of:
   A. Frequency distribution
   B. Range
   C. Measures of central tendency
   D. Measures of variability
10. What is the
    A. Mean?
    B. Median?
    C. Mode?
11. Why is the standard deviation frequently preferred rather than the average deviation?
12. Explain how grades are assigned using:
    A. Absolute criterion
    B. Relative criterion

13. Select a shop class and determine the grades using the absolute criterion. Using the same group determine the grades using a relative criterion.
14. Find the means and standard deviation for the students in the related class having the following scores on a test:
    98, 95, 93, 90, 87, 85, 85, 84, 80, 76, 72, 70, 66, 64, 61, 61, 59, 57, 49, 34, 30.
15. Assign grades to the above group. Justify your decisions.

REFERENCES

1. Harrisberger, Lee, "The Cram Exam Is a Sham," *ERM Newsletter,* Vol. 1, No. 3 (February, 1967).
2. From *Short-cut Statistics for Teacher-made Tests,* Copyright © 1960 by Educational Testing Service. All rights reserved. Second edition, 1964. Reprinted by permission.

Chapter 22 USING FEEDBACK TO IMPROVE ASSESSMENT OF PERFORMANCE

Improvement of the instrumentation and methodology of performance evaluation requires analysis after use. Such analysis identifies weaknesses which often reduce the validity and the reliability of the process. With the present emphasis on achievement of satisfactory results careful attention to feedback is vital to improvement of the evaluation as well as the instructional process.

The first step in the review of the evaluation process is to assess the results achieved in terms of the performance goals identified. Were the desired performance goals achieved satisfactorily? If not, why not? Were the performance goals identified as to difficulty for the level of instruction, caliber of students, and time allocated to the subject? Was the content of the course adequately taught?

Review of Results on Paper-and-Pencil Tests

After tests have been given, time devoted to study of the results is essential if improvement of the tests and the learning process is to be achieved. This process is often described by the term "item analysis."

A quick and simple method of item analysis is to have each student as the class reviews the test indicate by a show of hands whether or not the response to the item was correct. As the teacher calls out the number of the item, all students holding a paper having the item wrong raise the right hand. The teacher counts the number of hands and subtracts the number from the total number in the class to show how many had the item correct. The number correct can then be written opposite the number of the item on the chalkboard or on the master copy of the test paper. If a large number of students responded incorrectly discussion of the item should follow. In this way a simple but effective item analysis is combined with the review of the test. This provides a basis for reassessing the wording or content of the test item.

The same approach can be used to perform the "high–low" type of item analysis. The purpose of this method is to show the powers of discrimination of each item. Statistically the top 27 percent and the bottom 27 percent of the total scores are singled out for attention. The purpose of this method is to determine whether or not more high scoring than low scoring students correctly responded to the item. Instead of using the top and bottom 27 percent, this method can be used with any desired percentage. When working with a class of students it is more practical to divide the top 50 percent from the bottom 50 percent. The method consists of:

1. Score the items of the test—this may be by students or teacher.
2. Arrange papers in descending order of total scores.
3. Count down to the middle score—all papers above the middle score fall in the high group; those below in the low group.
4. Divide the students in the room into two groups.
5. Give one group all the high papers and give the other group the low papers.
6. For each numbered item of the test provide space on the chalkboard to write:

Item No. *No. of Highs* *No. of Lows* *No.* $H + L$ *No.* $H - L$

7. As the item number is called out, the number of highs that got the item right is counted and recorded; likewise, the number of lows that got the item right is counted and recorded. Then the total number right is determined by adding. The difference between the number of right in the high group and the number of right in the low group is found by subtraction. This difference indicates the power of discrimination of the item. The scorekeeper always calls the numbers in the sequence: high, low, sum, and difference (i.e., 10, 6, 16, 4).
8. The difference between the high group and the low group should be at least equal to 10 percent of the number in the class. In the

illustration with 16 students in the class the difference between the high and low should be at least 1.6, but, in the illustration the difference was 4. Therefore, that item is satisfactory on a high–low criterion of item analysis.

The simple process described in these steps is identified in statistical methodology as finding the "biserial correlation with total test." The biserial correlation with total test is approximately equal to three times the high–low difference for items that 25 percent to 75 percent of the students answered correctly, if the high–low difference is based on the high–low halves of the class. In the illustration with a class size of 16 and a high–low difference of 4, the high–low percentage is 25 percent and the biserial correlation is three times this or .75. The error is greater at the extremes of highs or lows. It is desirable to have the average biserial correlation above .4. Greater detail of this process is provided in *Short-cut Statistics for Teacher-made Tests.* (1)

The process of item analysis can be completed by the teacher in the routine manner by tallying the number of correct responses of each of the students who completed the test. Then the teacher can make the "biserial correlation with total test" determination. This is a tedious process and is rarely done by most busy classroom teachers. The short-cut method is highly satisfactory as a means of identifying which items are effectively discriminating in the test. The process is basically the same; however, when the students participate in the item analysis it serves as a learning experience, can be done satisfactorily in a few minutes, and eliminates the long, drawn-out process for the teacher.

The next step consists of revising the items that were not working in the test and then using the test again.

Review of Results of Manipulative-Performance Tests

It is essential to provide the same type of analysis for manipulative-performance tests as for paper-and-pencil tests. The same process can be effectively used. An item analysis of each of the items in the test can be effectively made working with the students in the class as outlined in the previous section. Review of the items missed by the larger number of students provides an excellent learning experience for the students. In the case of manipulative-performance tests a repeat demonstration of the correct operation of tools or equipment may be highly desirable. The demonstration may be given by the instructor; but, in many cases, it can be given very effectively by one of the students who performed the operation correctly on the test. The shift of focus from the instructor to a member of the peer group provides a wholesome stimulus for learning for the students who failed to perform according to the established standard.

If items used on the manipulative-performance test were defective or did not properly discriminate, revision of the test item should take place prior to giving the test again. If the items were good but students needed more opportunity to apply the knowledges acquired to produce desired skill and adequate levels of confidence to perform effectively under the pressures of a test condition, consideration should be given to strengthening this aspect of the training program.

Review of Observations of Shop and Lab Work-in-Progress

Improved powers of observation, concentrating on specific elements for evaluation, and objective recording are key factors in observation. At the end of the course or at periodic intervals, time should be taken for a comprehensive review of the procedures used and the results achieved. A case study approach using a daily diary is helpful. Review of the daily diary provides a basis for improvement of the system. The critical incident technique is another helpful method to strengthen evaluation by observation.

A list of criteria of characteristics to be checked in observing students at work in the shop or laboratory can be helpful and reduce the work of recording observations.

Comparison of the observations with final outcomes is a step of the feedback process which may help to strengthen the instructional program.

Review of Completed Project

The ultimate test of ability to perform is reflected in the success or failure of the completed project. Correlation of achievement of standards with the instructional program is possible through evaluation of the project itself. Feedback reflects achievement of quality and/or quantity standards. Inherent in such evaluation is consideration of the techniques employed, equipment and tools used, as well as the skills and knowledges of the students.

PRODUCT EVALUATION

The product referred to is the student who is the product of the educational program. The success or failure of the student under actual conditions of employment can only be determined after the student has secured a position and his work has been evaluated by the employer. This takes considerable time. It also requires a follow-up procedure with the employer and the former student.

Summary

To have a closed-loop system of instruction, it is essential to connect the evaluation of performance of the student, both process and product, with the total program of instruction. The process consists of three fundamental steps:

1. Identification of performance components on the actual payroll job or cluster of payroll jobs for which vocational education is provided.
2. Formulation of the performance goals and construction of the instructional program.
3. Evaluation of the performance in relation to the performance goals and securing feedback for modification of the instructional program, if and as needed.

QUESTIONS AND ACTIVITIES

1. Why is feedback essential for the improvement of education?
2. How is feedback secured in the case of process evaluation? Of product evaluation?
3. What is the purpose of the item analysis of a test?
4. Can the method be applied to any type of test? Explain.
5. Compare two methods of making an item analysis. Which of the two is simpler and more rapid to complete?
6. What are the advantages of the high–low item analysis?
7. Explain the procedure for conducting a high–low item analysis.
8. What impact may result from item analysis?
9. How can the powers of observation be improved as a result of analysis?
10. What is the difference between process and product evaluation?
11. How is feedback secured for product evaluation?
12. Construct a simple questionnaire for securing from the student information concerning success or failure on the job.

13. Develop a questionnaire for securing information from the employers relative to the former students' performance on the job.
14. In addition to the questionnaire method, how may the instructor obtain feedback for product evaluation?
15. How can feedback be used to help future students?

REFERENCES

1. *Short-cut Statistics for Teacher-made Tests.* Princeton, N.J.: Educational Testing Service, 1960, pp. 3–6.

Part V TEACHING PROCESS

Chapter 23 EMPLOYING APPROPRIATE TEACHING METHODOLOGY

Learning may be classified under three headings: action, reaction, and interaction. The teacher as the manager of the learning environment selects the method of teaching and learning for a particular lesson or part of a lesson. The method selected influences the effectiveness of teaching and the ease of learning.

It is the purpose of this chapter to briefly discuss methods of teaching appropriate for teaching related and applied instruction. Instructors try to vary the method and the approach to best meet the needs of the content being presented and the students in the class.

Action learning involves the student in actual performance of tasks. This method is most commonly employed in connection with shop and laboratory activities. It is also used to some extent in teaching related and applied classes in a classroom situation.

Reaction learning is a process of students reacting to the presentations of the instructor or information source in which the student is involved. This method often is identified with the lecture method of teaching. The activities of the students are often limited to taking notes and attempting to absorb the content which is being presented.

Interaction learning involves students in verbal exchanges with the instructor and other students. The conference method is an illustration of interaction learning. Interaction learning methods are usually more effective than reaction methods of learning.

Action Learning

Action learning is applicable to related and applied classes as well as to shop and laboratory classes. Some of the most effective action learning grows out of demonstrations.

DEMONSTRATION

In a demonstration the technique or method is shown either by the teacher or by a selected student. Demonstrations are usually more effective if these steps are followed:

1. Determine the performance goals of the demonstration in advance.
2. Plan how to arouse the interest of the student.
3. List and secure needed tools, equipment, materials, and supplies.
4. Identify the teaching points to be covered during the demonstration.
5. Plan the method of presentation.
6. Determine the assignment to the student following the demonstration.
7. Select the additional materials needed for the students' involvement.
8. Develop a suitable performance test or paper-and-pencil test.

COMMITTEE WORK

Division of the class into smaller groups or committees permits greater participation by each individual in the group. It also provides opportunities for development of student leadership.

A common practice is for the instructor to select the temporary or permanent chairman. The group selects the secretary. For each individual to participate actively the group should be composed of between five and ten students. The chairman should be selected for his ability to guide the discussion and achieve the objectives previously identified. The secretary records the important aspects of the discussion. The secretary may also be requested to give the report of the committee deliberations to the total group as well as prepare a written report.

Committee work is most effective if:

1. Fundamental concepts and pertinent information have been presented in advance to all students.
2. Guidelines have been developed to aid the work of the committee.
3. The assignment is clearly delineated.
4. The chairman is selected for his ability to lead a group and for his understanding of the task assigned.

5. Each member of the committee is motivated to participate actively and in a meaningful manner.

INDIVIDUALIZED INSTRUCTION

As each individual is given a task or assignment different from others in the group, responsibility for completion rests on that individual. Give the student responsibility. Make him responsible for his own learning. But, help him carry the responsibility, if necessary!

The "software" approach to individualized instruction may be developed through the use of assignment sheets, operation sheets, information sheets, job sheets, and job plan sheets.

The "hardware" approach to individualized instruction grows out of one person using a teaching aid for self-study. This may be a super-8 motion picture projector, a slide projector, a tape recorder, a videotape recorder, or a computer station.

A good aid for individualized instruction needs to:

1. Explain the abstract idea and show clearly essential relationships to relevant applications.
2. Be clearly visible and understandable.
3. Follow a logical step-by-step sequence.
4. Use sketches, drawings, and charts effectively to aid the learner.
5. Use sketches, drawings, and other graphics in correct proportions to each other.
6. Have explanatory messages wherever the interpretation is in question.
7. Provide a summary or review for the learner.
8. Stimulate desire for application and make application easy.

PROGRAMMED INSTRUCTION

Programmed instruction is another educational technique which may use either hardware, software, or both. In programmed instruction the material is organized and presented in such a form as to permit the student to progress at his own rate. Self-testing is an important feature of programmed instruction. Emphasis is placed on the accomplishment of specified objectives by each student. The system includes built-in features of self-evaluation and automatic reinforcement of essential learning processes. Programmed materials are available in the form of programmed textbooks, commercially prepared instructional sheets, and laboratory manuals. Some teachers have developed their own programmed materials.

INSTRUCTIONAL RESOURCES

Resources for learning and teaching increase the effectiveness and ease of the process. These resources need to be selected for the kind of learning and used effectively with the students. The competent instructor is familiar with and uses effectively each of these tools of teaching. A few of the most common and most helpful are mentioned here.

Chalkboard. While a comparatively inexpensive aid, a chalkboard can be extremely helpful. It becomes a temporary reference for the student and serves as a reminder to the teacher. Words, outlines, and sketches placed on the chalkboard serve as helps to the student.

Each instructor should constantly strive to apply good technique in working at the chalkboard. A few ideas to remember include:

1. Identify your course and yourself on the top lefthand corner of the chalkboard.
2. Begin writing on the chalkboard at the extreme left.
3. Write large enough for students to read.
4. Try to maintain words in a straight line.
5. Use the panels of the chalkboard as pages of a book.
6. Erase the chalkboard neatly before beginning to write.
7. Form the letters distinctly.

Textbooks and reference books. Textbooks and reference books are invaluable aids to learning and teaching. Selection of books appropriate to the level of the student is essential. Books that are too elementary are as undesirable as those that are too advanced.

Students in class may need to refer to the text for additional information or to a handbook for specific data. Supplementation of textbooks with reference manuals, other textbooks, and trade and technical journals broadens the understanding and aids in the solution of many specific problems. Encourage students to use the resources of the library and to become familiar with indexes and guides to reference works.

Periodicals. Recent changes and new innovations are discussed in periodicals. Each student should have the opportunity to become familiar with the most important periodicals in his (or her) field. Stimulation to read periodicals results from emphasis placed on the importance of keeping abreast of new developments.

A class period designated for reports by each student on a new development is an excellent way of encouraging students to read periodicals. Some teachers designate one or two periods a month for such reports.

Charts. In many vocational fields charts are available from the manufacturers or distributors. These are frequently suitable for display on the wall or a chart rack. Charts suitable for class use should be acquired and properly used to supplement the instructional process.

Film strips. Often appropriate film strips are available commercially or from manufacturers. A library of film strips selected to coincide with the objectives of the course is tremendously helpful to the instructor. These provide a sense of realism and authenticity to the instructional program which should not be ignored.

Slides. Many instructors have taken slides of processes, products, or other suitable subjects which support the objectives of the course. This is a comparatively inexpensive and very effective visual aid.

Films. Rental or free films are available in many occupational areas. Sources of information on available films include:

1. H. W. Wilson's *Educational Film Guide* and *Filmstrip Guide.*
2. Educators Progress Service's *Educator's Guide to Free Films.*
3. *Educator's Guide to Free Slidefilms.*
4. U.S. Office of Education, *A Directory of 3,660 16mm Film Libraries.*
5. University film libraries.

Tapes. Commercially prepared and teacher developed tapes have been used extensively by some teachers of related and applied classes.

Audiovisual tapes. Audiovisual tapes are now developed and used as aids to teaching and learning in many vocational and technical schools. Great progress has been made in the improvement of the equipment and technique necessary for use of audiovisual tapes in classroom situations.

Closed-circuit television. Projection television provides improved opportunities for students to view demonstrations. It also enables large groups in removed places to have ringside seats at the site of action.

Computerized instruction. Computer-assisted instruction (CAI) is an instructional technique that uses a computer and a simple, English-based programming language (course writer) for the development and presentation of educational materials. This gives the student the equivalent of a private tutor in the specialized field. Developmental, remedial, corrective, and enriching materials may be included in the instructional materials provided by CAI.

Reaction Learning

It has been said that "telling is not teaching." Telling may or may not be teaching, depending on how the telling takes place and how the student reacts to the telling. Telling is frequently described as lecturing.

In the lecture method the instructor presents the information through the spoken word. Students react to the words that have been spoken. In every situation there is a sender (instructor), receiver (student), and message. Achievement depends upon the effectiveness of both the sender and the receiver as well as the nature and stimulation of the message.

Lectures may be formal presentations to large groups or short informal presentations to very small groups. While this method is often described as the most ineffective method of teaching, it is used in many situations. The degree of effectiveness is very dependent upon how well the instructor has prepared himself and the student. Some lectures are very boring to the student while others are extremely stimulating. In using the lecture method, the most effective results will be achieved if:

1. The lecture is carefully planned and well organized.
2. The subject is of interest to the students.
3. Motivation is introduced at the very beginning to catch the attention of the student. Combat the students' "ho-hum" attitude.
4. Cases are used—usually students are interested in the human interest story and illustrations or experiences.
5. The treatment of the subject is "pitched" at the level of the students. Define and explain unfamiliar terms. Avoid talking over the heads of the students.
6. The speaker maintains eye contact with the audience. The most effective lecturers have developed a technique of communication with the audience through the eyes. This provides a sensitivity to the audience reaction.
7. The talk is short and focused on the subject.
8. The room conditions are comfortable with adequate fresh air to avoid drowsiness.
9. The lecturer gives the listeners specific points to listen for and react to.
10. The lecture is followed by:
    a. Question–and–answer session.
    b. Group discussion session.
    c. Demonstration period.
    d. Visual aids and discussion period.
    e. Symposium or round-table discussion.
11. The main points are summarized prior to the conclusion of the presentation.
12. The speaker uses his (or her) voice effectively, with variation in tone, tempo, and intensity.

Critical parts of the lecture are the introduction and the conclusion. The attention of the students must be secured through the opening remarks; maintained by challenging, meaningful, and dynamically presented content; and wrapped up into a neat package in the final conclusion.

Interaction Learning

Interaction is desirable in teaching related and applied classes. The successful teacher involves the student in the learning activities. The most skillful teachers use many different kinds of student interaction. These kinds of activities constitute the effective teacher's "bag of tricks." Selection of an appropriate kind of interaction for the learning problem is important. Once the involvement is underway, then the task is to manage the involved individuals and their treatment of the problems, questions, issues, or topics to produce the most significant use of time and resources.

Let us look at some of the more useful interaction learning methods.

DISCUSSION

The discussion method provides opportunities for individuals in the class to participate actively. Questions may be asked or comments made to clarify the ideas that have been expressed. The experiences of many individuals may be used effectively to aid the members of the class. Discussion can usually be used to stimulate interest of individuals who otherwise might have very little concern about the topic which is under consideration.

The discussion provides a more democratic setting than that of the lecture. If desired a group decision could be developed. However, the instructor must recognize that even group consensus can be incorrect, inappropriate, or undesirable.

Economical use of time results after both students and instructors have used this method enough to recognize how to:

1. Select appropriate topics for discussion.
2. Analyze the subject.
3. Delimit the subject and restrict the discussion to pertinent comments.
4. Combine concepts of many individuals into a meaningful whole.
5. Summarize the vital points relative to the topic.
6. Use this method effectively together with other methods of instruction.

Discussion may take several forms. It may involve the entire class in an open discussion of a topic or question presented by the instructor. This may follow or precede a lecture or be entirely independent of a lecture. The discussion may involve a selected number of the class in an initial presentation as in the case of a panel discussion, symposium, debate, or group interview. Following the initial presentation other members of the class may be invited to participate.

PROBLEM SOLVING

Problem solving is often part of the group discussion. The total class as well as individuals can be effectively incorporated in the solution of problems. The brainstorming technique may be used as an introduction to problem solving. The most logical approach to problem solving requires teachers and students to:

1. Recognize the problem.
2. Define the problem.
3. Gather or present data pertinent to the problem.
4. Analyze the data presented.
5. Formulate alternative solutions.
6. Weigh the various possible solutions and select the best one.
7. Test the merits of the solution selected.
8. Evaluate the success or failure of the solution for the problem.

The problem-solving method is an excellent tool to stimulate student participation. Group action rather than individual action is used to produce the results. This method can be used in any classroom, laboratory, shop, or teaching situation regardless of the subject area.

Application of the following suggestions will help ensure good results:

1. The problem should be relevant for the group and the subject.
2. Each member of the group must understand the problem and agree to a definition of the problem.
3. In gathering the data list as many causes of the problem as possible.
4. Identify as many possible solutions as possible. List those that may seem irrelevant.
5. In analyzing the solutions, discard—at the suggestion of the group—the least desirable solutions first.
6. After group discussion and selection of the best possible solution, the next step is action.
7. In determining a course of action the group should deal with such questions as who, what, where, when, and how.
8. While most of the evaluation will occur toward the end of the process, it should be a constant and ongoing process.

PANEL DISCUSSION

The panel discussion is an exchange of ideas by selected participants on a problem, issue, topic, or question. This is an approach which brings together a number of individuals who possess differing points of view. Each panelist should be qualified to make a worthwhile contribution. The chairman has the responsibility of ensuring that the panelists discuss the selected topic. He may prod them with questions or cut them off when the discussion becomes irrelevant.

In order to produce an effective panel discussion, it is important to:

1. Develop a detailed discussion plan.
2. Make use of the expertise of each panel member.
3. Adapt the plan to the personnel.
4. Select a panel discussion leader who not only understands the subject but has the ability to bring out the concepts from the panel members.

Instructional Management

The instructor in the classroom, shop, or laboratory performs a number of management functions. He manages and has responsibilities for the resources of instruction and the subjects of instruction, the students.

The human resource is our most valuable resource. From the time the individual first enters a formal schoolroom situation teachers are in a major role responsible for skills, knowledges, habits, and attitudes developed by the student.

As the manager of the learning environment and the learning resources the teacher is accountable and must be held accountable for much of the failure as well as given credit for much of the success that results from the formal educational experience of students.

There are many elements worthy of consideration relative to this topic. Effective instruction requires:

1. Responsible behavior on the part of the students as well as the teacher. Discipline, in a democratic manner, is essential in every work situation and should be required of every person.
2. Correct and safe use of tools and equipment. Every teacher must teach safety in relationship to the activities of the class. A good worker is a safe worker.
3. Use of resources. Careful planning to conserve resources is an important part of the learning process. Waste in school should not be tolerated, any more than waste in industry, of the materials of production.
4. Time. Time is one of our most precious commodities. Teach students to use time effectively. Strive to produce or achieve

some increment of accomplishment during every period. Even mistakes can be used as learning devices. Moments gone are gone forever.

5. Build teamwork. The vocational and technical class provides excellent opportunities for democracy in action. Use this opportunity to teach cooperation, teamwork, respect for the rights of others, and other desirable goals of our society. Integration of social concepts with the functions of the world of work provides the most effective methods of accomplishing the social goals of education.
6. Practice what you teach! As the teacher, do what you want the students to do. Your example is more important than what you tell the students. "What you do may speak so loudly that the students may not hear what you say!"
7. Treat your students in a friendly but not in a familiar manner. Be fair and impartial to all. Demonstrate patience and understanding in your teaching.
8. As a teacher you are also a counselor. Learn to listen as well as to talk. Develop the powers of keen observation.
9. Be professional in your appearance. Seek to instill the spirit of professionalism in your students.
10. Education is a business. As a manager of the educational environment and resources, conduct it as a business enterprise. But, there is a human side to enterprise. Let the human side of enterprise be reflected in your example and your treatment of the students.

Summary

Three classifications of learning are action, reaction, and interaction.

Types of action learning that involve the students in performance of tasks include demonstrations, committee work, and individualized instruction.

Reaction learning occurs when the student mainly listens and reacts, as in the case of formal or informal lectures.

Interaction learning involves the students in an exchange of ideas. This may occur as discussion, problem solving, panel discussions, etc.

Proper use of instructional resources expedites learning and teaching whether action, reaction, or interaction learning is involved. The instructor is a manager. He manages the processes for teaching, the resources for learning, and, to some degree, the recipient, the student.

Effective teachers recognize the role of education as a business enterprise while being constantly aware of the human factors as the purposes for establishing and continuing the enterprise.

QUESTIONS AND ACTIVITIES

1. Compare the role of the student in the following types of learning:
   A. Action
   B. Reaction
   C. Interaction
2. Explain the functions and responsibilities of the teacher for each type of learning:
   A. Action
   B. Reaction
   C. Interaction
3. How can lectures be made most effective?
4. What are the advantages and disadvantages as learning methods of:
   A. Group discussions
   B. Problem solving
   C. Panel discussions
   D. Demonstrations
   E. Committee work
5. When should the instructor employ large group instruction? Small group instruction? Individualized instruction?
6. Why should money be invested in instructional aids?
7. What instructional aids do you feel produce the greatest "return" for the amount of money invested in teaching related or applied courses?
8. Discuss the changing role of the instructor.
9. Why can the term "instructional manager" be correctly applied to the teacher? Explain.
10. How can teachers become more effective managers of the human side of enterprise?

Chapter 24 TEACHING PROCESS: INDIVIDUAL, SMALL GROUP, AND LARGE GROUP

The process of teaching varies with the conditions, mode of instruction, and the size of the group. The purpose of the process of teaching is to provide the most effective learning environment for the students. All learning is individual.

While the teaching process is carried out in various ways with various types of content, age group, and size of group, there are some common characteristics in the total process applicable to all situations. In order to effectively set the stage for learning and managing the learning environment, the teacher must apply a planned, integrated system which carefully considers what the teacher needs to do and what the learner must do to be successful.

Working from an established curriculum with identified performance goals the teacher plans, prepares, presents, and evaluates through feedback the success of the learners' achievement. Discussion of each of these steps was provided in Chapter 14, pp. 224–228.

Working within the framework of the environment for learning, the student prepares, applies, and demonstrates ability to perform skills or mastery of concepts and knowledges identified as required learning activities.

Both the teacher and the students make adaptations to the learning environment, i.e., space provided, learning aids, and nature of the learning assignment. Opportunities for learning and teaching will vary with the nature of the teaching process and the size of the student group. Through adjustment of media and careful consideration of the techniques to be used, the learning process may be quite effective with large groups under some conditions. Under other conditions a small group or individualized instruction is essential for effective learning.

The most common system of instruction is described as the four-step method. The four-step method consists of preparation, presentation, application, and testing or follow-up. This method may be complete in one lesson or it may extend over several lessons.

Preparation of the Teacher

The preparation step consists of two parts. The first part is the preparation of the instructor to teach. During this step the teacher identifies the performance goals for the lesson and plans how to achieve these goals and how to know whether or not the goals have been achieved.

Also during this step the teacher plans appropriate subject matter content and how to teach the content. The instructor assembles all the materials, supplies, and equipment for the lesson.

One of the most important preparations of the teacher is that preparation which makes the content to be presented meaningful and relevant to the student. Consolidating the relationships of the content fragments of the curriculum under immediate consideration with those elements previously taught and those parts still to be considered is necessary, if the student is to grasp the significance of the entire educational program in relationship to the objectives of the curriculum.

Preparation of the Student

If the student is not ready to learn, he will accomplish very little. A basic law of psychology tells us that a student learns best when he is ready to learn. This is the law of readiness. The students may not be self-motivated to learn. It is the responsibility of the teacher to gain the attention of the student and arouse his desire to learn. Desire on the part of the student may be a latent rather than an active drive. The teacher may stimulate the desire of the student to learn by:

1. Relating today's lesson to a known fact, example, or past experience.
2. Demonstrating a significant principle or interesting application.
3. Asking questions to stimulate thought and student participation.
4. Showing posters, charts, models, samples, or drawings of the work or aspects related to it.
5. Using samples of other students' work to stimulate interest.
6. Capitalizing on the curiosity and desire of students to create in the area of content of the lesson.
7. Introducing specialists from the world of work to present the needs and desires of industry, business, or agriculture.
8. Taking field trips to industry, business, or agriculture to better show what modern industry is like and what the new employee will be expected to do and know when commencing entry-level employment.
9. Engendering suitable mind sets, attitudes, or moods which lead to a genuine interest in the subject matter content.
10. Explaining the specific objectives clearly and relating the desired performance goals to the outcomes of the course and the employment desires of the students.
11. Informing the students of success in learning. The success story of previous graduates is a tremendous motivational device. The success of members of the group on previous performance activities stimulates the desire to press forward. *Nothing succeeds like success.* Good salesmen tell success stories.
12. Testing achievement frequently. When students know where they stand relative to the identified objectives, they are more likely to be motivated to greater efforts.
13. Commending students for accomplishments when commendation is justified is an excellent stimulant to further growth.
14. Conferring with individual students may be essential to spark greater desire on the part of the individual and may result in motivation transmitted to the entire class.
15. Capitalizing on the students' interests.

Psychologists have frequently indicated that students usually work at a small percentage of potential capacity. Stimulation and motivation are necessary to bring out the potential capabilities of each individual. Motivation is closely related to morale. The good teacher is constantly concerned with building and maintaining the morale of the group at a high level. Meaningful participation in relevant activities closely related to known goals coupled with closely monitored progress toward the established goals strengthens the self-motivation of most individuals.

Presentation

In the presentation step, new concepts or applications are presented to the learner. The new elements of the lesson are introduced to the student. The method used to introduce the new elements may consist of a single technique or a combination of techniques common to the methodologies of teaching. The technique may consist of telling, showing, or appealing to a number of senses.

The teaching process may consist of demonstration, conference, discussion, lecture, problem solving or a combination of these with other methods.

Teaching aids in the form of charts, sketches, models, mock-ups, real objects, and similar devices may be used effectively during the presentation cycle of the lesson. Media suitable for large groups or mass media, i.e., closed-circuit television, may be employed. In other instances, media geared to individualized instruction may be used.

The success of the presentation step depends largely upon the instructor's familiarity with the subject and his ability to secure, produce, and use effectively instructional aids and teaching methods. Instruction should be carefully planned using the most effective aids and media available. Simple, clear, and meaningful instruction is usually most effective.

Each of the steps and key points needs to be stressed. Then, time must be given the students to absorb the content to which they have been exposed. The steps need to be placed in the most logical sequence from the simple to the more complex. The rate of presentation must be adjusted to the needs of the learners. Inject essential explanation, stressing the relationship of the "how" and the "why."

In the case of demonstrations show the correct way at normal rate. After this has been done, then slow down and go through the demonstration in slow motion. Answer questions and pause for the students to grasp the procedures. Reinforce your demonstration with repetition of essential steps and explanation of key points. Then, repeat the demonstration at the normal rate to provide students with a realistic sense of pacing of the job. Later, provide opportunities for the student to perform "dry runs" under close supervision of the instructor, prior to actually commencing the operation or job.

In performing the presentation step, follow a logical, orderly development of the topic, but be flexible enough to adjust the instruction to the needs of the students. Use questions during the presentation to stimulate the students' thinking, to check understanding, and to develop class participation.

The lecture method is the poorest method of presentation. A combination of discussion with lecture or a conference approach is usually more desirable.

Individualized instruction is enhanced by use of "software" as well as "hardware." "Software" may consist of a series of operation sheets, information sheets, job sheets, job plan sheets, sketches or blueprints, and assignment sheets. Teaching machines or computerized instruction are "hardware" approaches to individualized instruction.

Application or Try-out Performance

The third step of the sequence is the application of knowledge or skill gained. This often takes the form of try-out application under the immediate supervision of the instructor. If the student cannot apply that which he (or she) has learned, the learning is unsatisfactory.

During this phase of the learning process the student tries to do what he has been shown or explain what he has learned. During this step the learner demonstrates his ability to do and begins the formation of correct habits and right attitudes toward the operation, job, or occupation. This is a critical aspect of the learning process. Sufficient time must be given to this phase of learning to produce a firm base for further growth. As the student applies that which he (or she) has learned and the instructor corrects, modifies, reinforces, or encourages the student, progress occurs. Repetition with conscious effort to improve is essential to develop confidence which is vital to effective performance on a payroll job.

As the learner begins to apply skill and knowledge to a new task, he may be slow and awkward. The instructor must be patient and provide encouragement to the student. During this period the learner must concentrate on how to perform the tasks even though the quality and quantity of effort are considerably less than will later be demanded of the student on entry-level jobs. Emphasize correct performance! If the student is performing incorrectly, stop the learner and reteach the correct procedure. Provide opportunities for the student to practice until he understands each step and recognizes each key point of the process.

Instructors may ask questions during this step related to: Why? What? Where? When? Who? How? While verbalizing the process may be helpful, it must not be interpreted as demonstration of the actual ability to perform the tasks of the job.

Test or Follow-up

During this step the student performs by himself and is checked according to established performance goals and standards of performance for the activity. The method employed depends upon the lesson or unit and the nature of the instructional process.

One method of checking may require the learners to respond to a question or a series of questions. These may be oral or written. A mechanized process assuring individualized response to each question by each person may be used. A student may be requested to summarize the important aspects of the lesson. Another student may be asked to indicate the most important elements learned from the lesson.

Another method of evaluating achievement of the learner consists of an object test or a manipulative-performance test. During this process the instructor observes and evaluates the process as well as the result of the test. In other cases the students may be assigned a job to be performed. If the learner is expected to perform the job without help he is in effect performing a "test" job.

Checking the results of the test, relating essential findings to the students, reteaching if necessary, and then providing additional opportunity to apply skills or knowledges are essential during the initial phase of the learning process. Retesting the student with the same or a similar job is highly desirable to determine accomplishment by the student and also to enable the student to progress to the next learning challenge. As the student progresses he builds confidence, which is essential for continued growth in the learning process. Refinement of performance, improved quality, and increased quantity usually result from repetition combined with challenging motivation. As the learner becomes more proficient, he will need to be checked less frequently by the instructor.

The follow-up or test is proof of performance. It is also evidence of the job of teaching. Such evaluation helps both the student and the teacher and results in economical use of time and the resources of learning.

Summary

The purpose of teaching is learning for the student(s). The most common teaching processes consist of demonstrations, conferences, lectures, problem-solving activities, or combinations of these methods. Media and teaching aids enhance learning. All learning is individual.

The teaching and learning process consists of four steps:

1. Preparation
2. Presentation
3. Application
4. Test

Motivation is an important part of preparing the student to learn. The resourceful teacher adapts the motivational technique to the interests and needs of the students.

The four steps of the teaching process may be completed in a lesson of short duration or one requiring several periods.

QUESTIONS AND ACTIVITIES

1. Why should the teacher be competent in many processes of teaching?
2. What processes are most commonly used with related classes? With shop classes?
3. What methods are more frequently used with individualized instruction? Small group instruction? Large group instruction?
4. What are the two aspects of the teacher's preparation? Explain.
5. Describe effective techniques to use in motivating students in related or applied classes.
6. List several common ways to make presentations effective.
7. Why is the application step essential to the learning process?
8. What is the difference between the application step and the test?
9. Prepare a lesson plan for a related lesson of short duration.
10. Prepare to teach the lesson planned in question 9 to a group of students. If possible videotape the lesson and study the rerun of the total lesson. If a video recorder is not available use a tape recorder and listen to the replay.
11. Evaluate your own teaching using the video recorder or the tape recorder.
12. Review the teaching of your peers and identify strong and weak points.

Chapter 25 VIEWING THE FUTURE

Vocational and technical education is at the crossroads! What will be accomplished in the future depends upon many factors. Students are more and more recognizing the rightful role of vocational education. Many parents still suffer from the illusion that "college" is the only answer for their children.

Vocational education is gaining more support in terms of both funding and moral support from both the federal and state units of government. Local boards of education are placing more emphasis upon vocational education.

Serious questions still exist relative to the priorities in vocational education. Individuals who are to teach must be competent in two fields. First, the field that they teach, and second, the teaching field. Too frequently in trade and industrial and also in technical education, individuals are hired to teach directly out of industry without any preparation for teaching other than experience in the occupation. This really isn't fair to the prospective teacher or the students that he is expected to teach. New teachers must have opportunities to acquire knowledge of and skill in the teaching process and how to organize the instructional and curriculum materials for teaching. Time should be provided for this type of preparation through formal courses, short institutes, or individualized instruction before the new teacher begins to teach.

Good Teachers

All teachers want to be good teachers. How are good teachers described? A good teacher has been described as having:

1. The ability to associate ideas and things. In vocational education it is more effective to go from things to ideas rather than from ideas to things. The good vocational and technical teacher starts with things and moves to theory only to support practical applications.
2. Interest and concern for the task at hand. He tries to work toward the upper limits of his ability. He also tries to get his students to work toward the upper limits of their abilities. Many individuals work at only 10 or 15 percent of their capacity. Most students who fail in vocational and technical education fail because of lack of motivation, not because of lack of ability. The good teacher recognizes this and adjusts the challenges, tasks, and goals to stimulate students. He (or she) builds an environment and builds the students' reputations to emphasize success. He gives the students responsibility and makes them a part of the educational enterprise. As active partners, he rewards and recognizes achievement in such a way as to provide even greater desire on the part of the students.
3. Drive and energy. He works hard and expects his students to work hard too. He also plays hard. He is "success oriented."
4. A sense of timing and a realistic recognition of essential rates of production, both for himself and for his students. He recognizes that time is a precious commodity which must be conserved.
5. The ability to take difficult concepts and principles and simplify them so that students master the difficult without realizing that the concepts are difficult. A poor teacher takes simple concepts and makes them difficult. A good teacher takes difficult concepts and makes them simple.

INDICATORS OF SUCCESS

It has been said that during successful days in school:

1. More time was spent approving students' actions than in disapproving the action of the students.
2. More time was spent in asking questions than in answering them.
3. At least one new idea was planted in the minds of the students.
4. At least one new thought was stimulated.
5. The teacher did more listening than talking.

6. Time was provided for students' discussion and active participation in learning activities.
7. Good humor was maintained throughout the learning activities.
8. Every student moved closer to the goal and recognized that he had made some forward progress.

The Teacher as a Professional

The teacher is a professional person. He acts and thinks as a professional. He treats others as the professional he is and in return expects to be treated in the same way. Often it has been said that some teachers are not very professional—this may be true. Each must strive to overcome the obstacles that tend to be barriers to responsible professionalism. A professional person:

1. Works effectively with a minimum of supervision.
2. Concentrates on the job and not on the clock.
3. Takes responsibility for his actions.
4. Strives constantly to improve himself and to keep abreast of new developments.
5. Seeks to constantly contribute to the welfare of his profession.
6. Participates actively in his professional organizations.
7. Respects the confidence of others.
8. Is loyal to his fellow workers and is discreet in handling confidential information.
9. Uses established channels for resolution of problems and grievances.
10. Meets his commitments and obligations to his students, his fellow teachers, his administration, and his employer.
11. Is proud of being a vocational or technical teacher and demonstrates his sincere commitment and dedication to all.

Teachers' Role in Building the Future

Never in the history of our nation have so many teachers worked with so many students in classes in vocational and technical education. More schools offering vocational and technical curriculums are in existence in the United States now than in any previous period. More new and different instructional programs are available to interested students than ever before. More funds from more different sources are now available than to any previous generation of teachers.

Progress has been made—but only the surface has been scratched. The challenge of the future is great. A breakthrough is now apparent for those who believe in vocational and technical education. If this is to serve mankind—as vocational and technical education can—renewed efforts must be made to:

1. Help more students to recognize the tremendous opportunities that await them in vocational education.
2. Instill in parents a realization that vocational education offers an excellent opportunity for jobs with satisfying qualities and good financial rewards for young people both for today and tomorrow.
3. Counsel more vocational guidance and other guidance personnel about vocational and technical education.
4. Enlist the aid of the peer group in communicating opportunities in vocational and technical education to other students.
5. Inform disadvantaged and handicapped about vocational and technical education.
6. Build the status of vocational and technical education. The crawling phase is over, now is the time to move with dignity and firmness on all fronts:
   A. Service
   B. Finance
   C. Public information
7. Provide a systems approach so necessary for an effective, integrated program of balanced vocational and technical curriculums geared to industries' need.
8. Validate that which is being taught in all vocational and technical classes.
9. Accept the challenge of accountability—welcome it! This will make vocational and technical education look good when the facts are known.
10. Build relevant learning experiences for students of all ages and groups who enter classes in vocational and technical education.

The future belongs to those who prepare for it!

INDEX